Bioarchaeology and Identity in the Americas

Bioarchaeological Interpretations of the Human Past:
Local, Regional, and Global Perspectives

UNIVERSITY PRESS OF FLORIDA

Florida A&M University, Tallahassee
Florida Atlantic University, Boca Raton
Florida Gulf Coast University, Ft. Myers
Florida International University, Miami
Florida State University, Tallahassee
New College of Florida, Sarasota
University of Central Florida, Orlando
University of Florida, Gainesville
University of North Florida, Jacksonville
University of South Florida, Tampa
University of West Florida, Pensacola

Bioarchaeological Interpretations of the Human Past:
Local, Regional, and Global Perspectives

Edited by Clark Spencer Larsen

This series examines the field of bioarchaeology, the study of human biological remains from archaeological settings. Focusing on the intersection between biology and behavior in the past, each volume will highlight important issues, such as biocultural perspectives on health, lifestyle and behavioral adaptation, biomechanical responses to key adaptive shifts in human history, dietary reconstruction and foodways, biodistance and population history, warfare and conflict, demography, social inequality, and environmental impacts on population.

Bioarchaeology and Identity in the Americas

EDITED BY KELLY J. KNUDSON
AND CHRISTOPHER M. STOJANOWSKI

Foreword by Clark Spencer Larsen

University Press of Florida
Gainesville/Tallahassee/Tampa/Boca Raton
Pensacola/Orlando/Miami/Jacksonville/Ft. Myers/Sarasota

Copyright 2009 by Kelly J. Knudson and Christopher M. Stojanowski
Printed in the United States of America on acid-free paper

First cloth printing, 2009
First paperback printing, 2010

Library of Congress Cataloging-in-Publication Data
Bioarchaeology and identity in the Americas / edited by Kelly J. Knudson and
Christopher M. Stojanowski ; foreword by Clark Spencer Larsen.
p. cm.—(Bioarchaeological interpretations of the human past : local, regional,
and global perspectives)
Includes bibliographical references and index.
ISBN 978-0-8130-3348-8 (alk. paper); ISBN 978-0-8130-3678-6 (pbk.)
   1. Indians—Anthropometry. 2. Indians—Population. 3. Ethnoarchaeology—
America. 4. Group identity—America. 5. Paleopathology—America. 6.
Paleoanthropology—America. I. Knudson, Kelly J. II. Stojanowski, Christopher
M. (Christopher Michael), 1973–
E59.A5B56 2009
305.897—dc22   2008045644

The University Press of Florida is the scholarly publishing agency for the State
University System of Florida, comprising Florida A&M University, Florida At-
lantic University, Florida Gulf Coast University, Florida International University,
Florida State University, New College of Florida, University of Central Florida,
University of Florida, University of North Florida, University of South Florida,
and University of West Florida.

University Press of Florida
15 Northwest 15th Street
Gainesville, FL 32611-2079
http://www.upf.com

This book is dedicated to our families.

Contents

Figures

Tables

Foreword

A person's identity is shaped by many factors, but political economy, social organization, and the history of society are at the top of the list. These inform social constructions and uses of sex and gender, age, access to food and other resources, and physical characteristics to understand a person's place and role in society. Rather than being a simple list of static attributes, a person's identity is dynamic, reshaping and reconfiguring. This creativity is what underlies identity and sense of social place within the community and society. As the contributions to this book so wonderfully illustrate, bioarchaeologists as social *and* biological scientists are well positioned to document, interpret, and contribute to understanding identity in the past—especially at the individual and community levels—because they study the person's remains (the biological) and the mortuary and broader archaeological context (the social) from which these remains derive. Moreover, bioarchaeology is well situated to study identity because of temporal context and the ability to assess change over time, along with the remarkable biological record left in the skeleton, including sex, age, health history, lifestyle, genetic history, and natal origin. Some excellent examples of identity and what the study of human remains can provide are presented in this book. In a number of settings worldwide, manipulation of the body during life in the form of cranial deformation and dental modification provides a kind of memory of social and ethnic identity.

The chapters in this book focus on the bioarchaeology of identity, but the common theme that adds to its coherency is Spanish colonization of the Americas and its profound impact on social and cultural behavior of native populations. In this regard, the study of mortuary behavior underscores the important point of just how malleable ethnic identity can be, especially when viewed over long stretches of time. On the other hand, some aspects of identity endure over many generations. For example, while treatment of the body at the time of death superficially appears to replicate European Christian practice in colonial settings—body in supine posture, hands folded on the chest—the presence of non-Christian, "native" material culture in association with the deceased is emblematic of the conservative nature of some aspects of social and religious practice and ideology in general (see Thomas 1988 for another example from North America).

It has long been understood that identity is accessible in the archaeological record via the study of human remains, especially with regard to inferences about gender and health (e.g., various authors in Grauer and Stuart-Macadam 1998). What makes this volume such an important advance, however, is the careful crafting of problems and application of new theory and method in developing an understanding of fundamental socially derived aspects of human behavior. The volume moves bioarchaeology and its practitioners from a place of study of human remains from archaeological settings to a broader understanding of these remains as though the populations they represent were alive today, as social beings and with all the complexities of identity that play such an important part in how social identity is constructed.

Clark Spencer Larsen
Series Editor

References Cited

Grauer, Anne L., and Patricia Stuart-Macadam, editors. 1998. *Sex and Gender in Paleo-pathological Perspective*. Cambridge: Cambridge University Press.

Thomas, David Hurst. 1988. "Saints and Soldiers at Santa Catalina: Hispanic Designs for Colonial America." In *The Recovery of Meaning in Historical Archaeology*, edited by M. P. Leone and P. B. Potter, 73–140. Washington, D.C.: Smithsonian Institution Press.

1

The Bioarchaeology of Identity

KELLY J. KNUDSON AND CHRISTOPHER M. STOJANOWSKI

The nation states of the last few hundred years have given us a historically unusual notion about identity: we feel that identity should have fixed borders within which similarity of language, customs, legal systems and physical types are quite different from those across the border. The fictive nature of the nation has recently become apparent and has helped us see that identity always exists in a tension between fixity and flux.

Gosden 2004: 156

Identity is an extremely important focus of research in the social sciences and humanities today, despite its nebulous nature, which defies easy monolithic definition. At the most general level, identity refers to peoples' perceptions of themselves and how they relate to larger social phenomena that characterize their existences. Identity can be based on psychological affect as well as situational circumstance. More specifically, identity can have political, economic, or religious modes of expression and can be "ethnic" in orientation, nationalist, tribal, or linguistic. Identity also relates to an individual's sex and gender as well as social and biological age. It is the totality of these perceptions and ascriptions that constitutes an individual's cohesive sense of self and that informs actions and social constraints. Obviously the diffuse nature of the concept requires careful analytical reconstruction, and for this reason most contemporary research is focused on modern phenomena (see Brubaker and Cooper 2000; du Gay et al. 2000; Hutchinson and Smith 1996; Romanucci-Ross and De Vos 1995). Accessing data on social identities in the past is much more difficult, but crucial for understanding the historical dimensions of this important component of the human experience. Bioarchaeology is particularly well suited to address research on past identities because of the attendant time-depth offered by an archaeological chronological framework as well as the direct engagement of the physical body in the construction of social identity. Indeed, in both regards bioarchaeology offers an ideal approach to this topic. Despite this, previous bioarchaeological contributions to the study of identity have been limited—a lacuna this volume explicitly seeks to correct.

In the 30 years since the tenets of bioarchaeology were first outlined (Buikstra 1977), the field has witnessed substantial growth (Buikstra 2006a; Larsen 2006; Roberts 2006). Bioarchaeology has moved beyond its origins as subsidiary descriptive osteology (Buikstra and Beck 2006) and has more recently responded to critiques of basic operational assumptions, such as the population-death assemblage correspondence, the osteological paradox, and the French school's criticisms of paleodemography (Bocquet-Appel and Masset 1982; Cadien et al. 1974; Wood et al. 1992). The discipline has thus matured and emerged from these challenges in a more theoretically informed and methodologically robust form, as witnessed by the expansion of programs in the United States and abroad (e.g., Roberts 2006) and the breadth of new research areas to which bioarchaeologists are contributing. Although methodological refinement continues (e.g., Baker et al. 2005; Hoppa and Vaupel 2002), the broadening significance of contextual analysis of human remains reflects a certain comfort with our data and methods and discontent with purely descriptive and methodological studies as a *raison d'être* for our research programs (see Armelagos and Van Gerven 2003; Stojanowski and Buikstra 2005; Wright and Yoder 2003).

As the field sheds its status as a methodological research tool for the study of human remains from archaeological contexts, it is redefining its role as a theoretically informed discipline providing insights into ancient and modern communities as uniquely signaled in mortuary remains. As a materials-based empirical methodology (the "bio-") united with a social science inherently committed to understanding the human social world (the "-archaeology"), we move effectively between the social and biological sciences to address both models of social process and issues of a historical nature. As methodologies have improved and the number of practitioners has increased, research interests have also expanded. This is part of the normal progression of a discipline. Although known primarily for its contributions to prehistoric health, in particular health changes associated with the agricultural and post-Columbian transitions (Baker and Kealhofer 1996; Cohen and Armelagos 1984; Larsen 2001; Larsen and Milner 1994; Steckel and Rose 2002), recent work has moved beyond this focus and is pushing the field into new and challenging directions that fully utilize the benefits of bioarchaeological approaches. Part of this trend is to stress the importance of contextualization, a recurrent theme throughout several recent treatments (Buikstra and Beck 2006; Gowland and Knüsel 2006; Rakita et al. 2005; Sofaer 2006a).

In addition to the increasing focus on contextualizing data within an appropriate archaeological and historical framework, recent monograph-length treatments have also started to move in new research directions. For example, Lori Wright's (2006, based on her 1994 dissertation) recent publication on Maya diet, health, and status is framed within the broader study of state collapse. Wright (2006) clearly sees the breadth of implications that her data sources offer and uses health and dietary data based on pathology and isotope analyses to argue that political and social conflict, rather than environmental degradation, contributed to the demise of the Classic Maya in the Pasión region. This is a significant contribution to regional issues *and* to more general theories of societal collapse. Jean Guilaine and Jean Zammit (2005) focus on the general study of warfare and conflict, with prehistoric skeletal data playing a central, but not singular, role in their research. In addition, the number of regional bioarchaeological surveys is growing (e.g., Hutchinson 2002, 2003; Oxenham and Tayles 2006), as are synthetic multidisciplinary treatments of specific diseases and their social histories, which have important public health implications (e.g., Powell and Cook 2005; Roberts and Buikstra 2003).

The current volume continues this trend. Drawing heavily from social theory and historical ethnography, it presents a collection of papers that explicitly examine identity by linking regional historical and archaeological data with bioarchaeological approaches, including population genetic analyses, biogeochemical analyses of human tooth enamel and bone, analysis of mortuary contexts, and inferences based on body modifications. Volume contributors focus on identity in North, Central, and South American populations, predominantly dating from AD 500 until the colonial period. Chapters highlight the interplay of social phenomena and human biology in the New World to add a valuable biological component to our understanding of identity, which traditionally has been the purview of social theorists, ethnographers, and historical archaeologists.

A Brief History of Identities Research

Initial anthropological interest in social identity focused almost exclusively on ethnicity. The concepts of ethnicity and ethnic groups popular in the social sciences today are derived from Fredrik Barth's (1969a) influential book and the work of the Manchester School in the 1960s and 1970s (Cohen 1969, 1974; Cohen 1978; Gluckman 1958). These scholars emphasized the situational and highly mutable character of ethnicity, which can be manipulated and adopted in order to achieve desired outcomes even though membership in an ethnic group must be self-identified and readily distinguished by non-group members (Banks 1996; Barth 1969a, 1969b, 1969c; Cohen 1969, 1974; Cohen 1978; Gluckman 1958). As a vehicle of economic or political mobilization, ethnicity was seen as a useful analytical concept that broke away from the static notion of "tribe" and its attendant implications of primitive, bounded, and internally homogenous groups (Banks 1996; Jones 1997). The work of Barth and other social scientists was also a distinct departure from that which stressed the primordial and psychological nature of ethnic identity (De Vos 1975; Epstein 1978; Geertz 1963; Isaacs 1974), including its sociobiological implications (van den Berghe 1978, 1981). Since these initial explorations, practice theory has been applied to the use of ethnicity in anthropology in an attempt to explain how people come to embrace a shared social identity (Bentley 1987; Fisher and DiPaolo Loren 2003; Jones 1997; Yaeger 2000; Yelvington 1991).

Although initial anthropological interest in ethnic identity was immediate, the study of ethnicity in archaeology has occurred relatively recently. It remains controversial, in part because of the difficulty of defining ethnicity in a way that is identifiable in the archaeological record. Despite the difficulties inherent in using material remains to address something as situational and ephemeral as ethnicity, archaeologists have used ceramics, domestic architecture, lithics, basketry, textiles, food, body ornamentation, and mortuary ritual to distinguish one ethnic group from another (reviewed in Emberling 1997; Jones 1997). The most robust studies use multiple lines of evidence to strengthen their inferences. But the primary focus of investigation has typically been synchronic, following Barth's (1969b) concern with identifying a social group as "ethnic" in character and delimiting a field of interaction that forms the boundaries between different ethnic groups. Geoff Emberling (1997) notes that the clearest examples of ethnic boundaries and social differences are often found in ethnic enclaves, such as the immigrant barrios at Teotihuacan (e.g., Price et al. 2000; Rattray 1990). Even in the absence of clear ethnic enclaves, however, multiple lines of evidence can allow the archaeological study of ethnicity (Aldenderfer 1993; Brumfiel 1994; Erdosy 1995; Hill 1987; Kamp and Yoffee 1980; Larick 1991; Pollard 1994; Ponting 2002; Pyszczyk 1989).

In addition to the growing body of literature on the archaeological manifestations of ethnicity, there is an increasing recognition that an individual adopts any number of social, religious, and political identities that may coexist or may change over time (Díaz-Andreu et al. 2005; Meskell 2001; Schortman et al. 2001). For example, feminist contributions to identity research focus on the relationships between gender and sex and on the importance of studying gender as an identity (e.g., Arnold and Wicker 2001; Brumfiel 1992; Butler 1999; Conkey and Spector 1984; Díaz-Andreu 2005; Hays-Gilpin and Whitley 1998; Joyce 2000; Meskell 2001; Sweely 1999; Wright 1996). More recent research has also considered the role of archaeology in reconstructing the relationship between identity and modern political processes, and vice versa (Arnold 1999; Díaz-Andreu and Champion 1996; Dietler 1994; Gosden 2006; Higueras 1995; Kane 2003; Levy 2006; Meskell 2002; Shennan 1989; Tilley 2006). Despite the clear importance of these nonethnic identities to society and the individual, such complex and shifting identities are notoriously difficult to examine archaeologically. Archaeologists have attempted to elucidate social, religious, and/or political identities through, for example, mortuary artifact assemblages (Joyce 1999), residential architecture and artifact assemblages (Cecil 2004; Dietler and Herbich 1998; Hodder 1979; Janusek 2002, 2003a, 2003b; Pollard 1994; Schortman et al. 2001; Woodhouse-Beyer 1999), clothing (DiPaolo Loren 2003; Pollard 1994), and cranial and dental modification (Blom et al. 1998; Geller 2006; Hoshower et al. 1995; Torres-Rouff 2002, 2003). This brief sketch is hardly sufficient to cover the broad range of topics subsumed within the literature (additional information is available in chapter 2 by Buikstra and Scott, this volume).

The Role of Bioarchaeology in Identities Research

The bioarchaeology of identity as developed in this volume does not accept a unidimensional, vernacular connotation that implies geographic origin, measured by biodistance or biogeochemical analyses, or ancestral affiliation, through the much maligned classic taxonomic biodistance approaches. Rather, we define identities research as not about who people were or where they or their ancestors came from but about who they thought they were, how they advertised this identity to others, how others perceived it, and the resulting repercussions of this matrix of interpersonal and intersocietal relationships. More specifically, the questions that bioarchaeology addresses are as follows: How do identities begin and manifest both at the level of the individual, as in ensoulment and the creation of personal identity, and at the level of the community, as in ethnogenesis, ethnic emergence, and coalescence? How are markers of identity overtly displayed and manipulated across time and space, as in cranial modification or the use of specific material culture styles? How does the presence of multiple social identities, or social plurality, manifest itself? In what contexts does plurality lead to health disparity and under what circumstances is plurality less deleterious?

As previously discussed, bioarchaeology is ideally suited to the archaeological investigation of identity because of the unique time-depth and temporal perspective that our various research methodologies offer, frequently enhanced by the often immutable signatures of life written in the human skeleton (Buikstra and Beck 2006; Larsen 1997, 2002a, 2002b). In fact, Genevieve Fisher and Diana DiPaolo Loren (2003: 225) argue that "the representation and manipulation of the body is the most visual way to construct identity." Traditional bioarchaeological analyses such as sex and age determination are necessary for our understanding of identities based on sex, gender, and age (Díaz-Andreu 2005; Gowland 2006; Lucy 2005a, 2005b; Sofaer 2006a, 2006b; Walker and Cook 1998). In addition, when body modification is used to convey social or ethnic identity, bioarchaeologists have a unique perspective on a lifelong identifier such as cranial modification (Blom et al. 1998; Geller 2006; Hoshower et al. 1995; Torres-Rouff 2002, 2003). On the other hand, bioarchaeologists have an arsenal of techniques that can be used to determine biological relationships, such as biodistance analysis of cranial and dental nonmetric traits or ancient DNA analysis. In addition, a rapidly growing array of techniques can be used to determine life histories, such as strontium, lead, and oxygen isotope analyses of archaeological tooth enamel and bone. Finally, bioarchaeology provides a valuable, multigenerational perspective on the anthropological study of identity and ethnicity. Unlike ethnographic studies, which can infer identity formation and manipulation over the course of a few generations at best, bioarchaeological studies of identity elucidate long-term changes in identity formation and maintenance.

Another important contribution of bioarchaeological studies of identity is their ability to incorporate many lines of evidence to reconstruct identity over time and space. For example, Lynn Meskell (2001: 187–188) writes that "in our archaeological investigations we have concentrated on single-issue questions of identity, focusing singularly on gender or ethnicity, and have attempted to locate people from antiquity into a priori Western taxonomies: heterosexual/homosexual, male/female, elite/non-elite, etc. Archaeologists tend to concentrate on specific sets of issues that coalesce around topics like gender, age, or status, without interpolating other axes of identity, be they class, ethnicity, or sexual orientation, for example, because this has been seen as too vast or complex a project." In this volume, chapter contributors have attempted to address the complexity and variability of identity formation and manipulation by integrating multiple lines of bioarchaeological, archaeological, ethnohistorical, and ethnographic data.

The Structure of *Bioarchaeology and Identity in the Americas*

Given the broad range of topics subsumed within human social identity, it is necessary to define specific themes to be developed in this volume and certain parameters that we have established. The first parameter is geographic and temporal; we limit discussion to New World populations that primarily postdate AD 500. This focus is appropriate for several reasons. This volume complements recent monographs and edited volumes that focused primarily on bioarchaeology and social archaeology in Old World contexts (Gowland and Knüsel 2006; Sofaer 2006a), although the theoretical orientations, while overlapping, are not isomorphic, as developed by Buikstra and Scott in chapter 2 (this volume). However, we do not limit the contextual focus of contributors just to avoid regional duplication. Rather, concentrating our discussion on the New World minimizes a certain degree of variability in cultural trajectories, colonization models, time-depth, and biological variation. Although we do not divorce our contexts from general theory or assume a global, monolithic colonization model and subsequent indigenous colonial experience, it is clear that the contributors to this volume explicitly or implicitly engage their research questions through the lens of the Spanish colonial presence in the New World. Focus on more recent temporal contexts is also apparent, reflecting the importance of ethnohistorical texts and archaeological databases for enriching the breadth and nuances of the inferences provided by chapter contributors.

The second parameter is topical. Within the broad field of research on human identities we concentrate on two areas to which bioarchaeology can make the most important contributions: community identity and individual identity. This important distinction serves as the primary organizational structure in the chapters that follow, focusing on community identities in part I and individual identities in part II. Since the study of ethnic identities has the greatest time-depth in anthropology, we begin with community identity and ethnogenesis.

Part I: Community Identity and Ethnogenesis

In part I contributors explore two issues related to community identity in North and South American indigenous populations. The first is ethnogenesis, the process by which group distinctiveness is established and new ethnic identities emerge or are mobilized (Stojanowski, this volume; Nystrom, this volume; Sutter, this volume; Klaus and Tam Chang, this volume). The second is the more ephemeral concept of identity transcendence, in which aspects of a group's identity, primarily reflected in ideological inventories with nuanced reflection in the material record, transcend periods of social, political, or economic transition and which thereby establish cultural linkages that are not easily identified in colonial histories (Stojanowski, this volume; and Klaus and Tam Chang, this volume). We note that the concept of identity transcendence is distinct from the primordial essences associated with the psychological aspects of ethnic identification (e.g., Bromley 1974). The four chapters included in part I use two distinct methodological approaches: biodistance analysis in chapters 3–6 and mortuary analysis in chapter 6.

In chapter 3, "Bridging Histories: The Bioarchaeology of Identity in Postcontact Florida," Christopher M. Stojanowski explores community organization and identity during the late precontact, early mission, and late mission periods in Spanish La Florida (fourteenth through seventeenth centuries). Using dental metric data derived from seven different missions and numerous precontact sites analyzed using model-bound population genetic analyses, Stojanowski documents an increase in between-population genetic variability during the precontact/early mission period transition, followed by a rapid and significant decline in between-population genetic variation during the early/late mission period transition. These regional trends are interpreted in reference to ethnographic models of ethnogenetic change, which predict an initial separation phase in which preexisting ethnic sentiments lose saliency, followed by a liminal phase in which new intergroup connections are established. The biological data suggest similar trends in community relationships, indicating that ethnogenesis was ongoing among mission communities at the time of their destruction in the first decade of the eighteenth century. Biological data are complemented with historical information on local and global adaptive processes, which include population aggregation following demographic collapse and the tribalization of indigenous communities in response to distinct English and Spanish economic models that resulted in indigenous-indigenous conflict, such as slave raiding. The result of these historical processes was the genesis of a "pan–Spanish Indian" identity in La Florida, which is not reflected in contemporary historical accounts or colonial ethnonymy and is only suggested by population genetic signatures. Ethnohistoric, archaeological, and settlement data, combined with these bioarchaeological data, further suggest that the Florida Seminole have ancestral roots in this nascent Spanish Indian identity, an observation with important implications for the cultural patrimony of Florida's modern tribal communities.

In chapter 4, "The Reconstruction of Identity: A Case Study from Chachapoya, Peru," Kenneth C. Nystrom uses archaeological, ethnohistoric, and bioarchaeological data to examine the mechanisms responsible for Chachapoya ethnogenesis following the Inka conquest. Based on a series of population genetic analyses of craniometric data, Nystrom concludes that the area administratively called Chachapoya by the Inka was biologically heterogeneous yet united by limited long-range gene flow oriented in a north-south direction. These populations also experienced differential contact with extralocal populations. When combined with the ethnohistoric accounts of Chachapoya ethnic diversity during the preconquest period, as well as archaeological data on mortuary features and architectural design element variability, the biological data suggest that the Inka ignored significant ethnic and biological variation in their administrative structure but that Chachapoya ethnogenesis nonetheless commenced because of the common experience that highland populations shared in respect to their domination by the Inka state. Nystrom's synchronic approach to ethnogenesis moves beyond simple pattern recognition and boundary comparison exercises and explores the basis for ethnogenesis before the period of transition began. In this case, Chachapoya ethnogenesis, or the elevation of Chachapoya as a pan–northern highlands identity, likely only occurred because of Inka ascription and the dialectical, potentially ethnic-based structure that resulted from this process of objectification.

Bioarchaeological perspectives on ethnogenesis are also addressed by Richard C. Sutter in chapter 5, which is entitled "Post-Tiwanaku Ethnogenesis in the Coastal Moquegua Valley, Peru." Sutter presents a detailed archaeological analysis of ceramic design elements combined with a biodistance analysis of dental morphological variation of populations from the Moquegua and Azapa Valleys in southern Peru and northern Chile, spanning the Paleoindian through Late Intermediate Periods (ca. AD 1100–1400). Sutter is explicitly interested in the processes responsible for the ethnogenesis of the coastal Chiribaya after the Middle Horizon (ca. AD 500–1100) collapse of the Tiwanaku polity (see also Knudson and Blom, this volume). Using design matrices, Sutter argues for a two-stage diaspora model for Chiribaya ethnogenesis in which both coastal indigenous and post–Tiwanaku collapse middle Moquegua Valley colonists congregated in the lower valley and ultimately formed the Chiribaya polity, based on economic, mortuary, and ceramic data derived from archaeological sources. His model attributes Chiribaya ethnogenesis to economic co-dependency and the explicit shift in symbols of power after the collapse of the dominant political order represented at Tiwanaku.

These first three substantive contributions to *Bioarchaeology and Identity in the Americas* are united methodologically by a focus on heritable phenotypic variation and utilize standard and well-established techniques. The authors do not simply compare phenotypic distances between social units (such as archaeological sites), however, but elucidate the processes responsible for the ways in which these social units formed. A synchronic pattern recognition exercise is rejected in favor of temporal inferences about evolutionary process (Stojanowski, this volume; Sutter, this volume) or the relationship between a social construct with an associated ethnonym and a biological population in the formal sense of the word (Nystrom, this volume). These approaches are in stark contrast to many anthropological genetic analyses that compare patterns of biological and social variation (see also Stojanowski, this volume; and Nystrom, this volume). What these chapters demonstrate, and what bioarchaeological approaches offer in general, is the importance of temporal structure in a research sampling design. For biodistance analyses this time-depth allows assessment of evolutionary process alongside historical and social trajectories of change, and the quality of inference is considerably greater than for a strictly evolutionary focus invoking interpretations solely based on mechanisms such as adaptation (selection), population size (genetic drift), or similarity (gene flow).

In addition, chapters 3, 4, and 5 need to be understood in their broadest intellectual traditions. For example, both Stojanowski (this volume) and Nystrom (this volume) examine ethnogenesis within the context of expanding imperialist states and the power structure between the dominant political entity and the tribalized communities at the peripheries. Stojanowski (this volume) considers European imperialism in the context of postcontact North America, where the state and the subjugated are phenotypically and culturally quite distinct. In contrast, Nystrom (this volume) examines precontact ethnogenesis, where the cultural and biological traditions of the Inka state and the subjugated Chachapoya are not as distinct. The fact that this difference does not affect the outcome is intriguing and serves to normalize postcontact North American studies within the broader context of postcolonial theory.

Sutter's chapter is also about ethnogenesis but is decidedly different in orientation. Most importantly, Sutter considers ethnogenesis during a period of state collapse rather than imperialist expansion. This is an important distinction and part of the growing body of literature that addresses why states collapse (Diamond 2005) and what happens in the aftermath (Schwartz and Nichols 2006). Sutter's chapter fits nicely within this important literature, with clear contemporary applications and significance. Stojanowski (this volume) also considers human behavioral adaptation to a postcollapse environment; in this case the collapse is demographic and in reference to precontact chiefdom structure. Both Stojanowski and Sutter demonstrate how people navigate these difficult periods and emerge with new identities that incorporate aspects of their cultural heritage. This theme is discussed more fully by the final chapter in part I.

In chapter 6, "Surviving Contact: Biological Transformation, Burial, and Ethnogenesis in the Colonial Lambayeque Valley, North Coast of Peru," Haagen D. Klaus and Manuel E. Tam Chang present a reconstruction of postcolonial mission life for the Mochica population living in the village of Mórrope in the Lambayeque Valley, Peru. The authors use ethnohistoric and skeletal health data to demonstrate that the colonial Mochica suffered a declining quality of life as a sequella of demographic collapse. Despite this marginalized position, however, Klaus and Tam Chang identify a series of practices related to precontact Andean beliefs in soul-transfer and fertility rituals in the burials within the Catholic church; these include burial alignment, the inclusion of symbolic red mortuary textiles, the lighting of ritual fires, and an extended liminal period involving body manipulation. Since many of these practices were contrary to established Christian doctrine, what appeared to be standard Christian burials in a Catholic mission church are actually encoded with intense precontact ethnic symbolism. The authors interpret the inclusion of these symbols within mortuary contexts as a conscious emblem of resistance against the dominant colonial social order. Therefore, although the Mochica at Mórrope had every reason to succumb to ethnic extinction, precontact symbols uniting the present and the past were actively used to promote an ever-evolving Mochica identity. In this sense, elements of ethnic primordialism are evident, in which, "in the context of disorienting social change, people retire to their ethnic identity to meet emotional needs" (Yelvington 1991: 159).

One of the more intriguing interpretations presented by Klaus and Tam Chang (this volume) is the notion that the Mochica identity was defined by a conservative element manifest in their death rituals that transcended the Moche and Sicán collapse as well as Chimú and Inka conquest *before* Spanish conquest. Stojanowski (this volume) noted a similar cultural ideal that linked the eighteenth-century Seminole to their sixteenth-century forebears: that of resistance rather than accommodation. In both cases an identity "essence" is seen to transcend, and in the case of the Seminole actually implement, the ethnogenetic process. The importance for both the Mochica and the Seminole is the historical linkage afforded by the recognition of this enduring quality of identity. Such historical connections have modern implications for cultural patrimony and heritage policies and call into question facile parsing of ancestral connections using biological or cultural data alone. As the common theme of all four chapters in part I, the facts of ethnogenesis call into question simplistic parsing of indigenous histories while at the same time illuminating the psychological nature of colonial extirpations as part of the "struggle over a people's existence and their positioning within and against a general history of domination" (Hill 1996: 1).

Part II: Identity Formation and Manipulation at the Level of the Individual

In contrast to the chapters in part I, the authors in part II address identity formation and manipulation at the level of the individual, with a focus on Central and South American contexts. This individual-based approach highlights the unique ability of bioarchaeology to examine identity at radically different scales of analysis.

In chapter 7, "Cultural Embodiment and the Enigmatic Identity of the Lovers from Lamanai," Christine D. White, Fred J. Longstaffe, David M. Pendergast, and Jay Maxwell continue to explore the bioarchaeological manifestations of embodiment. After providing a thought-provoking overview of the theoretical approaches to embodiment and their relationship to bioarchaeology, the authors use bioarchaeological and archaeological data, including oxygen isotope ratios in human remains to identify geographic origins, to reconstruct the biological and social identities of three individuals buried together at the site of Lamanai in Belize. In addition to reconstructing the biological and social identities of the Lamanai lovers within their cultural context, White et al. promote the movement from traditional osteobiography to social biography for the purpose of better understanding social identity. More specifically, White et al. argue that "osteobiographical data such as individual histories of food consumption, disease experience, physical activity, and movement across the landscape should be put to greater use for inferring social identity."

In chapter 8, "Cranial Modification among the Maya: Absence of Evidence or Evidence of Absence," William N. Duncan effectively combines ethnographic and bioarchaeological data to understand Maya cranial modification practices from an ethnographic and linguistic model. He first notes that cranial deformation is a reflection of Maya embodiment and then questions how the large numbers of unmodified crania at Maya sites are typically interpreted. For example, Duncan asks: are individuals without modified crania somehow less fully embodied members of society than individuals with such modification? How might we reconcile the archaeological and ethnohistoric records? He argues that absence of cranial modification does not reflect a lack of embodiment. Using ethnographic and ethnohistoric data on Maya beliefs about the soul and ethnographic information on traditional child-rearing ceremonies, he demonstrates that children could potentially lose animating essences through their heads. Head binding was done during childhood rites of passage, the *héetz-méek'* ceremony, to seal off a potential window of soul loss. Duncan does not argue that variation in head shape was not intentionally created or was not a potentially meaningful social index in some Maya contexts. He does argue, however, that head binding was first and foremost an attempt to prevent such loss and did not necessarily have to result in a modified head shape to accomplish that goal. Although in this case cranial modification was often the result of these activities, it was not their most important goal. This runs counter to most interpretations of cranial modification through head binding and highlights the ever-present middle-range concern with linking observable archaeological residuals with the past behaviors responsible for those residues. In the case of Maya cranial modification, Duncan contends that the behavior itself carries much greater significance than its materially visible analogue.

In chapter 9, "The Complex Relationship between Tiwanaku Mortuary Identity and Geographic Origin in the South Central Andes," Kelly J. Knudson and Deborah E. Blom examine the creation and manipulation of social, political, and religious identities at the local level in response to interactions with larger and more powerful polities. They focus on the Tiwanaku polity of the South Central Andes, which exerted influence far beyond its Lake Titicaca Basin heartland during the Middle Horizon (ca. AD 500–1100). Knudson and Blom compare geographic origin through enamel and bone strontium isotope data with mortuary identity at Tiwanaku-affiliated sites in southern Peru and northern Chile. This chapter utilizes multiple lines of evidence, including geographic origin through biogeochemistry, genetic analyses through biodistance studies, cranial modification, mortuary artifacts, and burial treatments, to argue that the individuals buried at the Peruvian site of Chen Chen were in fact immigrants from the Tiwanaku heartland and articulated with the Tiwanaku polity as a colony or diaspora. The material culture, cranial modification data, and strontium isotope data at Chen Chen all point to a community-wide Tiwanaku social and political identity. In the San Pedro de Atacama oasis of northern Chile, however, Tiwanaku mortuary identity was *not* associated with migration from the Tiwanaku heartland, and oasis inhabitants may have actively manipulated their identity in response to contacts with the powerful Tiwanaku polity to the north. This unusual example of identity formation and manipulation in the archaeological record demonstrates the potential of multiple lines of bioarchaeological evidence to elucidate the complex relationships between material culture, geographic origin, and identity and highlights the dichotomies in self-perception, self-presentation, and objective ascription for interpreting identities in the past.

The San Pedro de Atacama oasis of northern Chile is also the location of the case study in chapter 10, "The Bodily Expression of Ethnic Identity: Head Shaping in the Chilean Atacama," by Christina Torres-Rouff. She examines cranial modification in the Chilean Atacama region to understand the construction and projection of social identities. Torres-Rouff argues that investigations of body modifications allow archaeologists to discern the individual and his or her agency in prehistory, a crucial element in exploring societal structures. In particular, since cranial modification must occur in the first few years of life, it imparts a permanent and socially meaningful marker of a child's individual identity. Through an analysis of 753 crania from seven cemeteries in San Pedro de Atacama, Chile, that date to AD 500–1500, Torres-Rouff (this volume) elucidates the social identities that Atacameños constructed and projected through time. During this period, Atacameños interacted with foreign powers and local exchange partners and witnessed substantial demographic shifts. In the Quitor and Coyo phases (AD 550–1000), two cemeteries reveal the use of this visible symbol by a portion of the population to affiliate themselves with foreign powers through similar styles of head shaping, concurrent with the maintenance of bodily expressions of local ethnic identity (see also Knudson and Blom, this volume). In contrast, crania from the subsequent Solor and Catarpe phases

(AD 1000–1500), a period of social and economic upheaval, demonstrate that head shaping was used to consolidate group identity. The reshaping of the head is a long and intimate process, and its presence in this group reflects social stability and the physical manifestation of long-lasting social identities.

A key theoretical approach in many of the chapters in part II is embodiment and its unique relationship with bioarchaeological approaches. Through analysis of the human skeleton, bioarchaeology can help elucidate the many ways in which the human body is used in complex political and social negotiations (e.g., Blom 2005; Fisher and DiPaolo Loren 2003; Joyce 2003, 2005; Sofaer 2006a). One aspect of embodiment that all contributors to part II utilized was cranial modification. While cranial modification could be used as a convenient and simplistic marker of individual identity, here the authors use multiple lines of evidence to provide much more complex and sophisticated understanding of the reasons behind cranial modification. For example, Duncan (this volume) argues that head binding during the Maya *héetz-méek'* ceremony was primarily a way to guard against soul loss. Torres-Rouff (this volume) argues that in the Andes some individuals used cranial modification to affiliate with foreign powers, while others used head shaping to consolidate group identity in later periods of social and economic upheaval.

Torres-Rouff's approach to investigating the use of cultural signifiers of identity in response to external and internal sociopolitical factors highlights another important theme in part II. Both Torres-Rouff (this volume) and Knudson and Blom (this volume) relate individual identity formation and manipulation to larger questions of state formation and expansion. This is a key counterpart to the chapters in part I, which examine identity formation and manipulation at the level of the population. Both approaches have contemporary relevance in a world where religious conflict and current concerns with nationalism and ethnic tribalism continue to figure prominently in world politics and, contrary to Marxist predictions, demonstrate no sign of abatement.

Finally, a third theme evident in the chapters in part II is the use of multiple lines of evidence to elucidate individual identity creation and manipulation. As in part I, all of the contributors to part II augment their bioarchaeological investigations with archaeological, ethnohistorical, and/or ethnographic data. For example, Torres-Rouff (this volume) examines the individual, the cranial modification itself, its distribution and typology, and the mortuary context to contribute to an understanding of the social identities that were being constructed and projected through time in San Pedro de Atacama. Similarly, White et al. (this volume) create sophisticated social biographies of the individuals buried in a unique mortuary context in the Maya site of Lamanai.

Conclusion

In conclusion, the chapters in this volume utilize social theory, ethnography, ethnohistory, and multiple lines of bioarchaeological and archaeological data to examine the diversity of manifestations of identity formation and manipulation in the past. By highlighting the relationship between social and biological identities, a nuanced and complex understanding of identity in the archaeological record can be achieved. In addition, the chapters share specific thematic and organizational elements. This is the design of the editors and reflects our vision for this volume. First, we explicitly minimize the presentation of data and methodologies in an attempt to shed bioarchaeology's status as a methodological tool. Second, chapter contributors present data sets and analyses generated through well-established bioarchaeological techniques but push their interpretations into new and sometimes challenging places. The importance of this volume lies in the new ways in which authors link their bioarchaeological data to broader problem orientations within the social sciences. Third, the chapters highlight the unique benefits of bioarchaeological data sets and present novel approaches to the past that are not available using other data sources. Fourth, and finally, the chapter contributors stress the importance of using multiple lines of evidence to effect more synthetic and nuanced interpretations of social identities in the archaeological record.

Acknowledgments

We thank the participants in this edited volume for stimulating and advancing our investigations of identity and bioarchaeology. We would also like to thank the students and faculty at Arizona State University, particularly in the Center for Bioarchaeological Research in the School of Human Evolution and Social Change, for their welcoming collegiality and thought-provoking scholarship.

References Cited

Aldenderfer, Mark S. 1993. *Domestic Architecture, Ethnicity, and Complementarity in the South-Central Andes.* Iowa City: University of Iowa Press.

Armelagos, George J., and Dennis P. Van Gerven. 2003. "A Century of Skeletal Biology and Paleopathology: Contrasts, Contradictions, and Conflicts." *American Anthropologist* 105: 53–64.

Arnold, Bettina. 1999. "The Contested Past." *Anthropology Today* 15: 1–4.

Arnold, Bettina, and Nancy L. Wicker, editors. 2001. *Gender and the Archaeology of Death.* Walnut Creek, Calif.: AltaMira Press.

Baker, Brenda J., Tosha L. Dupras, and Matthew W. Tocheri. 2005. *Osteology of Infants and Children.* College Station: Texas A&M University Press.

Baker, Brenda J., and Lisa Kealhofer, editors. 1996. *Bioarchaeology of Native American Adaptation in the Spanish Borderlands.* Gainesville: University Press of Florida.

Banks, Marcus. 1996. *Ethnicity: Anthropological Constructions.* London: Routledge.

Barth, Fredrik, editor. 1969a. *Ethnic Groups and Boundaries: The Social Organization of Culture Difference.* Boston: Little, Brown and Company.

———. 1969b. "Introduction." In *Ethnic Groups and Boundaries: The Social Organization of Culture Difference,* edited by F. Barth, 9–38. Boston: Little, Brown and Company.

———. 1969c. "Pathan Identity and Its Maintenance." In *Ethnic Groups and Boundaries: The Social Organization of Culture Difference,* edited by F. Barth, 117–134. Boston: Little, Brown and Company.

Bentley, G. Carter. 1987. "Ethnicity and Practice." *Comparative Studies in Society and History* 29: 24–55.

Blom, Deborah E. 2005. "Embodying Borders: Human Body Modification and Diversity in Tiwanaku Society." *Journal of Anthropological Archaeology* 24: 1–24.

Blom, Deborah E., Benedikt Hallgrímsson, Linda Keng, María C. Lozada Cerna, and Jane E. Buikstra. 1998. "Tiwanaku 'Colonization': Bioarchaeological Implications for Migration in the Moquegua Valley, Peru." *World Archaeology* 30: 238–261.

Bocquet-Appel, Jean-Pierre, and Claude Masset. 1982. "Farewell to Paleodemography." *Journal of Human Evolution* 11: 321–333.

Bromley, Yu. 1974. "The Term Ethnos and Its Definitions." In *Soviet Ethnology and Anthropology Today,* edited by Y. Bromley, 55–72. Hague: Mouton.

Brubaker, Rogers, and Frederick Cooper. 2000. "Beyond 'Identity.'" *Theory and Society* 29: 1–47.

Brumfiel, Elizabeth M. 1992. "Distinguished Lecture in Archeology: Breaking and Entering the Ecosystem—Gender, Class, and Faction Steal the Show." *American Anthropologist* 94: 551–567.

———. 1994. "Ethnic Groups and Political Development in Ancient Mexico." In *Factional Competition and Political Development in the New World,* edited by E. M. Brumfiel and J. W. Fox, 89–102. Cambridge: Cambridge University Press.

Buikstra, Jane E. 1977. "Biocultural Dimensions of Archeological Study: A Regional Perspective." In *Biocultural Adaptation in Prehistoric America,* edited by R. L. Blakely, 67–84. Athens: University of Georgia Press.

———. 2006a. "A Historical Introduction." In *Bioarchaeology: The Contextual Analysis of Human Remains*, edited by J. E. Buikstra and L. A. Beck, 7–25. New York: Academic Press.

———. 2006b. "Introduction to Section III: On to the 21st Century." In *Bioarchaeology: The Contextual Analysis of Human Remains*, edited by J. E. Buikstra and L. A. Beck, 347–357. New York: Academic Press.

———. 2006c. "Preface." In *Bioarchaeology: The Contextual Analysis of Human Remains*, edited by J. E. Buikstra and L. A. Beck, xvii–xx. New York: Academic Press.

Buikstra, Jane E., and Lane A. Beck, editors. 2006. *Bioarchaeology: The Contextual Analysis of Human Remains*. New York: Academic Press.

Butler, Judith. 1999. *Gender Trouble: Feminism and the Subversion of Identity*. New York: Routledge.

Cadien, James D., Edward F. Harris, William P. Jones, and Lawrence J. Mandarino. 1974. "Biological Lineages, Skeletal Populations and Microevolution." *Yearbook of Physical Anthropology* 18: 194–201.

Cecil, L. G. 2004. "Inductively Coupled Plasma Emission Spectroscopy and Postclassic Petén Slipped Pottery: An Examination of Pottery Wares, Social Identity and Trade." *Archaeometry* 46: 385–404.

Cohen, Abner. 1969. *Custom and Politics in Urban Africa: A Study of Hausa Migrants in Yoruba Towns*. London: Routledge.

———. 1974. *Urban Ethnicity*. London: Tavistock Publications.

Cohen, Mark N., and George J. Armelagos, editors. 1984. *Paleopathology at the Origins of Agriculture*. Orlando, Fla.: Academic Press.

Cohen, Ronald. 1978. "Ethnicity: Problem and Focus in Anthropology." *Annual Review of Anthropology* 7: 379–403.

Conkey, Margaret W., and Janet D. Spector. 1984. "Archaeology and the Study of Gender." *Advances in Archaeological Method and Theory* 7: 1–38.

De Vos, George. 1975. "Ethnic Pluralism: Conflict and Accommodation." In *Ethnic Identity: Cultural Continuities and Change*, edited by G. De Vos and L. Romanucci-Ross, 5–41. Palo Alto, Calif.: Mayfield.

Diamond, Jared. 2005. *Collapse: How Societies Choose to Fail or Succeed*. New York: Viking Penguin.

Díaz-Andreu, Margarita. 2005. "Gender Identity." In *The Archaeology of Identity: Approaches to Gender, Age, Status, Ethnicity, and Religion*, edited by M. Díaz-Andreu, S. Lucy, S. Babić, and D. N. Edwards, 13–42. London: Routledge.

Díaz-Andreu, Margarita, and Timothy Champion. 1996. *Nationalism and Archaeology in Europe*. Boulder: Westview Press.

Díaz-Andreu, Margarita, Sam Lucy, Staša Babić, and David N. Edwards. 2005. *The Archaeology of Identity: Approaches to Gender, Age, Status, Ethnicity, and Religion*. London: Routledge.

Dietler, Michael. 1994. "'Our Ancestors the Gauls': Archaeology, Ethnic Nationalism, and the Manipulation of Celtic Identity in Modern Europe." *American Anthropologist* 96: 584–605.

Dietler, Michael, and Ingrid Herbich. 1998. "*Habitus*, Techniques, Style: An Integrated

Approach to the Social Understanding of Material Culture and Boundaries." In *The Archaeology of Social Boundaries*, edited by M. T. Stark, 232–263. Washington, D.C.: Smithsonian Institution Press.

DiPaolo Loren, Diana. 2003. "Refashioning a Body Politic in Colonial Louisiana." *Cambridge Archaeological Journal* 13: 231–237.

du Gay, Paul, Jessica Evans, and Peter Redman, editors. 2000. *Identity: A Reader.* London: Sage Publications.

Emberling, Geoff. 1997. "Ethnicity in Complex Societies: Archaeological Perspectives." *Journal of Archaeological Research* 5: 295–344.

Epstein, Arnold L. 1978. *Ethos and Identity: Three Studies in Ethnicity.* London: Tavistock.

Erdosy, George. 1995. "Language, Material Culture and Ethnicity: Theoretical Perspectives." In *The Indo-Aryans of Ancient South Asia: Language, Material Culture and Ethnicity*, edited by G. Erdosy, 1–31. Berlin: W. DeGruyter.

Fisher, Genevieve, and Diana DiPaolo Loren. 2003. "Special Section, Embodying Identity in Archaeology: Introduction." *Cambridge Archaeological Journal* 13: 225–230.

Geertz, Clifford. 1963. "The Integrative Revolution: Primordial Sentiments and Civil Politics in the New States." In *Old Societies and New States: The Quest for Modernity in Asia and Africa*, edited by C. Geertz, 105–157. London: Free Press of Glencoe.

Geller, Pamela L. 2006. "Altering Identities: Body Modifications and the Pre-Columbian Maya." In *Social Archaeology of Funerary Remains*, edited by R. Gowland and C. Knüsel, 279–291. Oxford: Oxbow Books.

Gluckman, Max. 1958. *Analysis of a Social Situation in Modern Zululand.* Manchester: Manchester University Press.

Gosden, Chris. 2004. *Archaeology and Colonialism: Cultural Contact from 5000 BC to the Present.* Cambridge: Cambridge University Press.

———. 2006. "Postcolonial Archaeology: Issues of Culture, Identity, and Knowledge." In *Archaeological Theory Today*, edited by I. Hodder, 241–261. Cambridge: Polity Press.

Gowland, Rebecca. 2006. "Ageing the Past: Examining Age Identity from Funerary Evidence." In *Social Archaeology of Funerary Remains*, edited by R. Gowland and C. Knüsel, 143–154. Oxford: Oxbow Books.

Gowland, Rebecca, and Christopher Knüsel, editors. 2006. *Social Archaeology of Funeral Remains.* Oxford: Oxbow Books.

Guilaine, Jean, and Jean Zammit. 2005. *The Origins of War: Violence in Prehistory.* Malden, Mass.: Blackwell Publishing.

Hays-Gilpin, Kelley, and David S. Whitley, editors. 1998. *Reader in Gender Archaeology.* London: Routledge.

Higueras, Alvaro. 1995. "Archaeological Research in Peru: Its Contribution to National Identity and to the Peruvian Public." *Journal of the Steward Anthropological Society* 23: 391–407.

Hill, Jonathan D. 1996. "Introduction: Ethnogenesis in the Americas, 1492–1992." In *History, Power and Identity: Ethnogenesis in the Americas, 1492–1992*, edited by J. D. Hill, 1–19. Iowa City: University of Iowa Press.

Hill, Matthew W. 1987. "Ethnicity Lost? Ethnicity Gained?: Information Functions of

'African Ceramics' in West Africa and North America." In *Ethnicity and Culture: Proceedings of the Eighteenth Annual Conference of the Archaeological Association of the University of Calgary*, edited by R. Auger, 135–139. Calgary: University of Calgary, Archaeological Association.

Hodder, Ian. 1979. "Economic and Social Stress and Material Culture Patterning." *American Antiquity* 44: 446–454.

Hoppa, Robert D., and James W. Vaupel, editors. 2002. *Paleodemography: Age Distributions from Skeletal Samples*. Cambridge: Cambridge University Press.

Hoshower, Lisa M., Jane E. Buikstra, Paul S. Goldstein, and Ann D. Webster. 1995. "Artificial Cranial Deformation at the Omo M10 Site: A Tiwanaku Complex from the Moquegua Valley, Peru." *Latin American Antiquity* 6: 145–164.

Hutchinson, Dale L. 2002. *Foraging, Farming, and Coastal Biocultural Adaptation in Late Prehistoric North Carolina*. Gainesville: University Press of Florida.

———. 2003. *Bioarchaeology of the Florida Gulf Coast: Adaptation, Conflict, and Change*. Gainesville: University Press of Florida.

Hutchinson, John, and Anthony D. Smith, editors. 1996. *Ethnicity*. Oxford: Oxford University Press.

Isaacs, Harold R. 1974. "Basic Group Identity: The Idols of the Tribe." *Ethnicity* 1: 15–41.

Janusek, John Wayne. 2002. "Out of Many, One: Style and Social Boundaries in Tiwanaku." *Latin American Antiquity* 13: 35–61.

———. 2003a. "The Changing Face of Tiwanaku Residential Life: State and Local Identity in an Andean City." In *Tiwanaku and Its Hinterland: Archaeology and Paleoecology of an Andean Civilization: Volume 2, Urban and Rural Archaeology*, edited by A. L. Kolata, 264–295. Washington, D.C.: Smithsonian Institution Press.

———. 2003b. "Vessels, Time, and Society: Toward a Ceramic Chronology in the Tiwanaku Heartland." In *Tiwanaku and Its Hinterland: Archaeology and Paleoecology of an Andean Civilization: Volume 2, Urban and Rural Archaeology*, edited by A. L. Kolata, 30–94. Washington, D.C.: Smithsonian Institution Press.

Jones, Siân. 1997. *The Archaeology of Ethnicity: Constructing Identities in the Past and Present*. London and New York: Routledge.

Joyce, Rosemary A. 1999. "Social Dimensions of Pre-Classic Burials." In *Social Patterns in Pre-Classic Mesoamerica*, edited by D. C. Grove and R. A. Joyce, 15–48. Washington, D.C.: Dumbarton Oaks.

———. 2000. *Gender and Power in Prehispanic Mesoamerica*. Austin: University of Texas Press.

———. 2003. "Making Something of Herself: Embodiment in Life and Death at Playa de los Muertos, Honduras." *Cambridge Archaeological Journal* 13: 248–261.

———. 2005. "Archaeology of the Body." *Annual Review of Anthropology* 34: 139–158.

Kamp, Kathryn A., and Norman Yoffee. 1980. "Ethnicity in Ancient Western Asia during the Early Second Millennium B.C.: Archaeological Assessments and Ethnoarchaeological Prospectives." *Bulletin of the American Schools of Oriental Research* 237: 85–104.

Kane, Susan. 2003. *Politics of Archaeology and Identity in a Global Context*. Washington, D.C.: Archaeological Institute of America.

Larick, Roy. 1991. "Warriors and Blacksmiths: Mediating Ethnicity in East African Spears." *Journal of Anthropological Archaeology* 10: 299–331.

Larsen, Clark Spencer. 1997. *Bioarchaeology: Interpreting Behavior from the Human Skeleton.* Cambridge: Cambridge University Press.

———, editor. 2001. *Bioarchaeology of Spanish Florida: The Impact of Colonialism.* Gainesville: University Press of Florida.

———. 2002a. "Bare Bones Anthropology: The Bioarchaeology of Human Remains." In *Archaeology: Original Readings in Method and Practice,* edited by P. N. Peregrine, C. R. Ember, and M. Ember, 111–128. Upper Saddle River, N.J.: Prentice Hall.

———. 2002b. "Bioarchaeology: The Lives and Lifestyles of Past People." *Journal of Archaeological Research* 10: 119–166.

———. 2006. "The Changing Face of Bioarchaeology: An Interdisciplinary Science." In *Bioarchaeology: The Contextual Analysis of Human Remains,* edited by J. E. Buikstra and L. A. Beck, 359–374. New York: Academic Press.

Larsen, Clark Spencer, and George R. Milner, editors. 1994. *In the Wake of Contact: Biological Responses to Conquest.* New York: Wiley-Liss.

Levy, Janet E. 2006. "Prehistory, Identity, and Archaeological Representation in Nordic Museums." *American Anthropologist* 108: 135–147.

Lucy, Sam. 2005a. "The Archaeology of Age." In *The Archaeology of Identity: Approaches to Gender, Age, Status, Ethnicity, and Religion,* edited by M. Díaz-Andreu, S. Lucy, S. Babić, and D. N. Edwards, 43–66. London: Routledge.

———. 2005b. "Ethnic and Cultural Identities." In *The Archaeology of Identity: Approaches to Gender, Age, Status, Ethnicity, and Religion,* edited by M. Díaz-Andreu, S. Lucy, S. Babić, and D. N. Edwards, 86–109. London: Routledge.

Meskell, Lynn. 2001. "Archaeologies of Identity." In *Archaeological Theory Today,* edited by I. Hodder, 187–213. Cambridge: Polity Press.

———. 2002. "The Intersections of Identity and Politics in Archaeology." *Annual Review of Anthropology* 31: 279–301.

Oxenham, Marc, and Nancy Tayles, editors. 2006. *Bioarchaeology of Southeast Asia.* Cambridge: Cambridge University Press.

Pollard, Helen P. 1994. "Ethnicity and Political Control in a Complex Society: The Tarascan State of Prehispanic Mexico." In *Factional Competition and Political Development in the New World,* edited by E. M. Brumfiel and J. W. Fox, 79–88. Cambridge: Cambridge University Press.

Ponting, M. J. 2002. "Roman Military Copper-Alloy Artefacts from Israel: Questions of Organization and Ethnicity." *Archaeometry* 44: 555–571.

Powell, Mary L., and Della C. Cook, editors. 2005. *The Myth of Syphilis: The Natural History of Treponematosis in North America.* Gainesville: University Press of Florida.

Price, T. Douglas, Linda Manzanilla, and William D. Middleton. 2000. "Immigration and the Ancient City of Teotihuacan in Mexico: A Study Using Strontium Isotope Ratios in Human Bone and Teeth." *Journal of Archaeological Science* 27: 903–913.

Pyszczyk, Heinz W. 1989. "Consumption and Ethnicity: An Example from the Fur Trade in Western Canada." *Journal of Anthropological Archaeology* 8: 213–249.

Rakita, Gordon F. M., Jane E. Buikstra, Lane A. Beck, and Sloan Williams, editors. 2005.

Interacting with the Dead: Perspectives on Mortuary Archaeology for the New Millennium. Gainesville: University Press of Florida.

Rattray, Evelyn C. 1990. "The Identification of Ethnic Affiliation at the Merchant's Barrio, Teotihuacan." In *Ethnoarquelogía: Coloquio Bosch-Gimpera*, edited by Y. Sugiura and M. C. Serra, 113–138. Mexico City: Universidad Nacional Autónoma de México.

Roberts, Charlotte A. 2006. "A View from Afar: Bioarchaeology in Britain." In *Bioarchaeology: The Contextual Analysis of Human Remains*, edited by J. E. Buikstra and L. A. Beck, 417–439. New York: Academic Press.

Roberts, Charlotte A., and Jane E. Buikstra. 2003. *The Bioarchaeology of Tuberculosis: A Global View on a Reemerging Disease*. Gainesville: University Press of Florida.

Romanucci-Ross, Lola, and George De Vos, editors. 1995. *Ethic Identity: Creation, Conflict, and Accommodation*. Walnut Creek, Calif.: AltaMira Press.

Schortman, Edward M., Patricia A. Urban, and Marne Ausec. 2001. "Politics with Style: Identity Formation in Prehispanic Southeastern Mesoamerica." *American Anthropologist* 103: 312–330.

Schwartz, Glenn M., and John J. Nichols, editors. 2006. *After Collapse: The Regeneration of Complex Societies*. Tucson: University of Arizona Press.

Shennan, Stephen J., editor. 1989. *Archaeological Approaches to Cultural Identity*. London: Unwin Hyman.

Sofaer, Joanna. 2006a. *The Body as Material Culture: A Theoretical Osteoarchaeology*. Cambridge: Cambridge University Press.

———. 2006b. "Gender, Bioarchaeology and Human Ontogeny." In *Social Archaeology of Funerary Remains*, edited by R. Gowland and C. Knüsel, 155–167. Oxford: Oxbow Books.

Sokolovskii, Sergey, and Valery Tishkov. 1996. "Ethnicity." In *Encyclopedia of Social and Cultural Anthropology*, edited by A. Barnard and J. Spencer, 190–193. London: Routledge.

Steckel, Richard H., and Jerome C. Rose, editors. 2002. *The Backbone of History: Health and Nutrition in the Western Hemisphere*. Cambridge: Cambridge University Press.

Stojanowski, Christopher M., and Jane E. Buikstra. 2005. "Research Trends in Human Osteology: A Content Analysis of Papers Published in the *American Journal of Physical Anthropology*." *American Journal of Physical Anthropology* 128: 98–109.

Sweely, Tracy L. 1999. *Manifesting Power: Gender and the Interpretation of Power in Archaeology*. London: Routledge.

Tilley, Christopher. 2006. *Journal of Material Culture Special Issue: Landscape, Heritage, and Identity*. Volume 11.

Torres-Rouff, Christina. 2002. "Cranial Vault Modification and Ethnicity in Middle Horizon San Pedro de Atacama, Chile." *Current Anthropology* 43: 163–171.

———. 2003. "Shaping Identity: Cranial Vault Modification in the Pre-Columbian Andes." Doctoral dissertation, University of California, Santa Barbara, Department of Anthropology.

van den Berghe, Pierre. 1978. "Race and Ethnicity: A Sociobiological Perspective." *Ethnic and Racial Studies* 1: 401–411.

———. 1981. *The Ethnic Phenomenon*. New York: Elsevier.

Walker, Phillip L., and Della C. Cook. 1998. "Brief Communication: Gender and Sex: Vive la Difference." *American Journal of Physical Anthropology* 106: 255–259.

Wood, James W., George R. Milner, Henry C. Harpending, and Kenneth M. Weiss. 1992. "The Osteological Paradox: Problems of Inferring Prehistoric Health from Skeletal Samples." *Current Anthropology* 33: 343–370.

Woodhouse-Beyer, Katharine. 1999. "*Artels* and Identities: Gender, Power, and Russian America." In *Manifesting Power: Gender and the Interpretation of Power in Archaeology*, edited by T. L. Sweely, 129–154. London: Routledge.

Wright, Lori E. 1994. "The Sacrifice of the Earth? Diet, Health, and Inequality in the Pasión Maya Lowlands (Guatemala)." Doctoral dissertation, University of Chicago, Department of Anthropology.

———. 2006. *Diet, Health, and Status among the Pasión Maya: A Reappraisal of the Collapse.* Vanderbilt Institute of Mesoamerican Archaeology Series. Nashville, Tenn.: Vanderbilt University Press.

Wright, Lori E., and Cassady J. Yoder. 2003. "Recent Progress in Bioarchaeology: Approaches to the Osteological Paradox." *Journal of Archaeological Research* 11: 43–70.

Wright, Rita P. 1996. *Gender and Archaeology.* Philadelphia: University of Pennsylvania Press.

Yaeger, Jason. 2000. "The Social Construction of Communities in the Classic Maya Countryside: Strategies of Affiliation in Western Belize." In *The Archaeology of Communities: A New World Perspective*, edited by M. A. Canuto and J. Yaeger, 123–142. London: Routledge.

Yelvington, Kevin A. 1991. "Ethnicity as Practice? A Comment on Bentley." *Comparative Studies in Society and History* 33: 158–168.

2

Key Concepts in Identity Studies

JANE E. BUIKSTRA AND RACHEL E. SCOTT

For the issue is really whether one can actually have an archaeology that is *not* concerned with identity.

Insoll 2007: 1

[W]e need to recognize that our bodies are not purely socially constituted—biology obviously plays a major factor, as well! . . . Yet there sometimes seems to be within the archaeology of identities an emphasis upon forgetting the prosaic, but equally important foundational rudiments such as biology, in favour of the more popularly perceived social theoretical elements. Moreover the empirical body from which adequate interpretation and theory are generated in pursuing past identities must also not be neglected; otherwise there is a danger that empty shells are created.

Insoll 2007: 4

Another interesting direction of recent mortuary studies has been to focus on the individual and on the emotive. . . . Most interesting here, however, is that nowhere in these published papers or examples did anyone suggest that such studies could be improved if they were done in conjunction with a physical anthropologist. Who better to assess and determine details about an individual?

Goldstein 2006: 380

During the closing decades of the twentieth century, *identity* emerged as a key construct within the social sciences and humanities, as scholars discussed and debated topics such as ethnicity, ethnogenesis, religious identity, and gender. Anthropologists entered the arena, followed by archaeologists, who have added a temporal dimension, largely from studies of material culture and architecture. As the impressive papers in this volume indicate, bioarchaeological perspectives also have much to contribute to investigations of identity, both of communities and of individuals.

Bioarchaeologists have only very recently focused upon identity. Why might that be? As a distinct discipline, bioarchaeology faces its own inherent intellectual and methodological challenges. Having to grapple with distinguishing between related constructs such as sex/gender or ethnicity/genetic heritage is a challenging task, requiring carefully considering multiple lines of evidence. For example, the developmental age categories commonly reported in osteological studies may not necessarily be those recognized as significant in the lives of past peoples. Not all disease is manifest in human remains, limiting inferences about infirmity, and defining the coupled social and individual constructs of disability is even more daunting. Privileged social status may be reflected in advantaged health, but not unequivocally, as in obesity among high-ranking individuals and lead poisoning among slave-owners. Religion is expressed most clearly, but not unambiguously, in mortuary contexts. Ritual sacrifices and manipulated remains are also informative, but they rarely occur. These are a few of the interpretative challenges that bioarchaeologists must face if they are to incorporate the physical body into studies of identity. As both Timothy Insoll (2007: 4) and Lynne Goldstein (2006: 380) underscore, however, the physical *and* the social body should be studied together.

A further challenge lies in theoretical differences between those who "study bones" and those who investigate archaeological contexts. Scholars working in the United Kingdom have emphasized such distinctions, which are glossed as the Science/Theory divide by Rebecca Gowland and Christopher Knüsel (2006b), who cite the differences between those engaged in scientific inquiry and those pursuing the interpretative archaeologies—a science/humanities dichotomy. Joanna Sofaer Derevenski (1997a, 1997b; Sofaer 2006a, 2006b) also highlights the tensions between the biological study of bones and the archaeological investigation of contexts. Instead of engaging in theoretical debates, American bioarchaeologists have fought to prevent their data and analyses from being relegated to appendixes of site reports, rather than assuming a prominent role in interpretations of the past (Buikstra 1991).

Terminology also poses a problem for scholars addressing identity in the past *or* present. Many authors fail to define terms, use them inconsistently, or adopt meanings (implicitly or explicitly) that differ from common usage. We address examples of this ambiguity in the concluding section of this chapter.

Despite such issues, bioarchaeologists are beginning to address identity in some interesting and creative ways. This volume is one such effort, explicitly designed to employ multiple lines of evidence, including historical and archaeological contexts, and to incorporate social theory. Given the theoretical and contextual focus that unites the following chapters, we first discuss the individual dimensions of identity commonly examined in the human sciences and then turn to a brief review of key concepts used in the study of identity. As the volume focuses upon American examples, we direct our discussion to the Western Hemisphere. We also venture across the Atlantic to the United Kingdom, where bioarchaeologists (or osteoarchaeologists) are also addressing issues of identity in a contextualized manner, but with less emphasis upon combining multiple lines of evidence in a problem-driven bioarchaeology. We underscore that the form of bioarchaeology advocated here is broadly based, including a range of topics, from mortuary (or funerary) archaeology to paleopathology.

Dimensions of Identity

While, as emphasized later, we favor a multidimensional approach to the study of identity, many archaeological and bioarchaeological studies tend to focus upon single dimensions. These commonly include religion, status, ethnicity, gender, age (especially childhood), and disability. Gender and age are most commonly combined (Sofaer Derevenski 1997a, 1997b), as in life-course analysis (Gilchrist 2000, 2004) or osteobiography (Robb 2002; Saul 1972; Saul and Saul 1989). With the exception of life-course analysis, we consider these dimensions separately here. Even so, we fully endorse Lynn Meskell's (2001: 188) call "to break the boundaries of identity categories themselves, blurring the crucial domains of identity formation, be they based on gender, sexuality, kin, politics, religion, or social systems." Identities are socially constructed, situational, and fluid.

Within archaeology, as within anthropology generally, certain dimensions of identity have a longer history of study than others. Of those selected here, religion, social status, and ethnicity have considerable time-depth, while gender, age, and disability are of more recent interest.

The study of *religion* is as old as anthropology itself (Frazier 1890) and remained significant throughout the twentieth century (Bloch 1992; Durkheim 1915; Eliade 1987 [1957]; Evans-Pritchard 1974 [1956]; Geertz 1973; Malinowski 1954; Rappaport 1968, 1999; Turner 1995 [1969]; and Wallace 1966). Archaeologists influenced by the "new" archaeology turned away from the topic as they pursued processual goals, frequently regional studies of settlement systems. The subject has reemerged, however, in the postmodern era of postprocessual (or processual-plus [Hegmon 2003]) theorizing (e.g., Insoll 2001, 2004). Aspects of religion and worldview, as noted above, can be read from the structure of mortuary sites and their contents as well as from other specialized activity areas and aspects of material culture. Processed human remains (with prepared mummies being an obvious example), along with other forms of body manipulation such as sacrifices and the decoration of human bones by painting or carving, are also frequently interpreted in light of religious principles (e.g., Rakita et al. 2005; Rakita and Buikstra 2008; and Tiesler and Cucina 2007).

Religious identity, along with social and political identities, is the subject of Knudson and Blom's contribution (this volume), which examines the impact of the Tiwanaku polity in remote regions of southern Peru and northern Chile. The strength of combining skeletal data with a nuanced consideration of funerary contexts is obvious in their work and in that of other contributors. Bone chemistry, biodistance data, cranial modification, funerary accompaniment, and cemetery structure combine to establish the nature of individual identities for immigrants and local residents far from the Tiwanaku heartland.

The investigation of *social status* differences also has a long history in mortuary studies and is appropriately based in differential treatment of the dead, forms of body modification, and even dietary differences (as noted in part II of this volume). During the late nineteenth and early twentieth centuries, most archaeologists assumed an isomorphic relationship between tomb complexity, grave wealth, and the social status of the deceased. Early twentieth century critiques, such as Alfred Kroeber's (1927) cross-cultural study, dampened enthusiasm for funerary archaeology. It was revitalized in the 1970s (Brown 1971), however, and continued in the Americas, even in the wake of Ian Hodder's (1982) critique, which was extended by workers such as Daniel Miller and Christopher Tilley (1984; see also Hodder 1984; Parker Pearson 1982, 2000; Shanks and Tilley 1982; Tilley 1984). Regional approaches tended to dominate late twentieth century discussions (Beck 1995; Chapman et al. 1981; Rakita et al. 2005), with social status retaining prominence, though interpretations were more nuanced than in previous decades.

In the United Kingdom, as noted by Margarita Díaz-Andreu and Sam Lucy (2005: 8), the impact of the postprocessual critique was more extreme, with status and religion being "the worst treated by postprocessual archaeologists." At least one article appears on each subject, however, in the recent Gowland and Knüsel (2006a) edited volume entitled *Social Archaeology of Funerary Remains*. Julie Bond and Fay Worley (2006) examine the roles of animals in Anglo-Saxon graves, suggesting that some may have served to represent totems and worldview. Social differentiation is addressed through stable isotope studies of diet by Janet Montgomery and Jane Evans (2006). Staša Babić (2005), in his historical treatment of status differentiation, also stresses the utility of dietary inferences along with habitual postures in reading status, noting that the postures have been of most interest in gender studies. Thus have physical remains entered identity discourse about social status in the United Kingdom.

Ethnicity has long been visible in the study of archaeological cultures, although not in its contemporary sense. An explicit assumption had been that differences in material culture assemblages could be used to identify distinctive ethnic groups. As Díaz-Andreu and Lucy (2005) note, a new perspective developed after scholars such as Stephen Shennan (1978) and Ian Hodder (1978) began to question the degree to which group identities coincided with material culture distinctions. Rather than material culture being a passive product, it became viewed as active in creating and maintaining ethnic (and other) distinctions. Similarly, people as actors assumed prominence in archaeological explanations. The fluidity of ethnicity was emphasized, as well as its situational nature, *sensu* Frederick Barth's (1969) edited volume entitled *Ethnic Groups and Boundaries*. Today archaeologists and bioarchaeologists struggle with the definition of ethnic groups in the past. Bioarchaeologists have tended to emphasize inherited skeletal and dental features, although, as we shall see below, other attributes are also being used.

As recognized by Nystrom in chapter 4 (this volume), Siân Jones (1997) has been a prominent figure in recent archaeological treatments of ethnicity. She explicitly defines ethnic identity as "that aspect of a person's self-conceptualization which results from identification with a broader group in opposition to others on the basis of perceived cultural differentiation and/or common descent" (Jones 1997: xiii). Her emphasis upon the "dynamic and historically contingent nature of ethnic identity in the past *and* in the present" (Jones 1997: 14) is a politically charged statement relevant to the inalienable nature of identities and territories. Jones (1997) thus implies that it is very difficult to trace ethnicities over time because of their fluid nature.

Jones (2007: 48) proposes a practice theory of ethnicity, borrowing Pierre Bourdieu's (1977) concept of *habitus* as "durable, often subliminal, dispositions towards certain perceptions and practices (such as those relating to the sexual division of labour, morality, tastes, and so on) . . . which are inculcated into an individual's sense of self at an early age." Because *habitus* includes repeated and regular activities, this knowledge can literally become embodied as repetitive actions shape the individual's body (Sofaer 2006a, 2006b). Jones (1997, 2007) cautions, however, that *habitus* and ethnicity are not the same. It can be argued, in fact, that *habitus* encodes a number of dimensions of identity, ethnicity being only one. The challenge then is how to distinguish between them, if that is a research goal. Even so, the study of *habitus* features embedded in the skeleton, such as musculoskeletal markers, as advocated by Sofaer (2006a, 2006b) for gender studies, may prove a useful line of evidence.

As noted by Lucy (2005b), studies of ethnicity fell out of favor during the post–World War II period in most European countries and in the United States. Interestingly there are no studies expressly of ethnicity in Gowland and Knüsel (2006a). By contrast, Stojanowski (this volume), Nystrom (this volume), and Sutter (this volume) address issues of ethnogenesis, displayed against a tapestry of inherited features, ethnohistoric sources, and archaeological contexts. All three chapter authors clearly understand, as does Lucy (2005b), that inherited features are not isomorphic with ethnicity but that they can inform when considered against contextual data. Ethnogenesis is also the subject of Klaus and Tam Chang's presentation in chapter 6 (this volume), but in the context of historic documentation, health data, and funerary rituals. Torres-Rouff (this volume) explores ethnic identity in her consideration of cranial modification as a means of embodying identity. She argues that distinctive forms of modification were used as a means of aligning residents of the site of San Pedro de Atacama with foreign powers such as the Tiwanaku polity, all while maintaining other material signals of local identity. This impressive display of hybrid political identities, also documented by Knudson and Blom (this volume), disappears after the fall of the Tiwanaku state, when cranial form embodies local political solidarity.

Pam Crabtree (1990) proposes that dietary differences may be useful in identifying ethnic differences. Other archaeological studies of ethnicity and diet include those of James Barrett et al. (2001), Charles Cheek and Amy Friedlander (1990), and Elizabeth Scott (2001). Dietary history is one of the attributes considered by White et al. in chapter 7 (this volume) in their discussion of social identity formations at the ancient Maya site of Lamanai in Belize. Both White et al. (this volume) and Knudson and Blom (this volume) employ other isotopic signatures (those of residence) to distinguish between immigrants and local residents as part of their multifaceted studies of identity.

Gender studies in archaeology are frequently informed by feminist theory, which is characterized by three waves of development (Gilchrist 1999). The first is associated with the suffragist movement of the late nineteenth and early twentieth centuries. The second, which had a more direct influence upon archaeology, began in the 1960s and inspired seminal articles, such as that by Margaret Conkey and Janet Spector (1984), and the significant volume edited by Joan Gero and Margaret Conkey (1991).

Conkey and Spector (1984), for example, illustrate a number of instances of androcentrism in archaeological interpretations, including the explanation of grave goods from funerary contexts. Citing Howard Winters' (1968) research on Midcontinental Archaic mortuary materials, they note that grinding pestles interred with women were considered tools used by the women for food-processing, but those interred with men were interpreted as items manufactured (rather than used) by the deceased. Foreign goods associated with men reflected their trading activities; with women, they were simply possessions. The mysterious presence of atlatls with both sexes led Winters (1968) to speculate about Amazons and Boudiccas. Clearly such interpretations involved preconceptions about gender-specific roles. More recently, contributors to a volume edited by Bettina Arnold and Nancy Wicker (2001) abundantly illustrate the issues that must be considered in developing a gendered perspective based upon objects from funerary contexts.

While the second feminist wave focused on individual equality at a personal level and on identifying the basis for women's oppression at an intellectual level, a third wave developed during the 1990s that was heavily influenced by postmodernism. Emphasis was less on individual rights and more upon the definition of differences between sexualities, ethnicities, and social classes. Thus the focus broadened from women to differences in general. In archaeology this has played out in terms of the study of gender differences, sexualities, and age identities (Gilchrist 1994, 1999, 2004).

Following definitions developed by Barbara Voss and Robert Schmidt (2000), Díaz-Andreu (2005: 14) asserts that "[g]ender can be defined as an individual's self-identification and the identification by others to a specific gender category on grounds of their culturally perceived sexual difference. The concept of gender is related but not equivalent to that of sex. Sex refers to the physical and genetic elements of the body related to reproduction, including genitalia, chromosomal and hormonal distinctions and reproductive organs." A related concept, sexuality, is seen as socially constructed and includes all kinds of sexual relations, such as sexual activities, eroticism, sexual identities, sexual meanings, and sexual politics. Schmidt and Voss's (2000) book is one of the few archaeological treatments of sexuality. Thus we have three related categories: (1) gender: social/cultural, (2) sex: biological, and (3) sexuality: social/cultural. To complicate matters further, queer theorists would collapse the categories of sex and gender, arguing that both are cultural constructs (Butler 1990; Meskell 1999). As Roberta Gilchrist (1999: 14) observes, this creates a problem for bioarchaeologists, who at least feel secure in being accurate in assigning biological sex to skeletal remains as much as 90 percent of the time.

Assuming that bioarchaeologists should *not* combine or confuse sex and gender, as Phillip Walker and Della Collins Cook (1998) argue, then how does the bioarchaeologist engender the past? Chris Meiklejohn et al. (2000: 227) reasonably point out that "[i]f biological defined 'sex' can be ambiguous, the culturally defined 'gender' must be a minefield." They do, however, proceed to develop a gendered perspective on the Mesolithic in the western Baltic through the study of skeletons, associated grave goods, and burial types (Meiklejohn et al. 2000). They conclude that since skeletally sexed males and females are buried with distinctly different grave goods, Mesolithic society was heavily gendered. This is a fairly standard approach to interpreting gendered roles from funerary contexts, though it is not without its bioarchaeological critics (Sofaer 2006a). Most archaeological studies of gender based on funerary contexts (such as Arnold and Wicker 2001) continue to use skeletal sex and age-at-death as the only reported physical parameters, rather than incorporating a broader range of individual attributes. This is typical of many group-based studies, which contrast with life-course (Gilchrist 2000, 2004) and osteobiographic approaches (Robb 2002) that combine gender and age identities.

In a pioneering study that both distinguished sex and gender analytically and also integrated age and gender identities, Sofaer Derevenski (1997a) investigated the Copper Age site of Tiszapolgár-Basatanya, Hungary. She compellingly argues that contemporary archaeological practices, which base the gendering of artifacts solely upon osteological analyses of sex and also exclude juveniles because they cannot be skeletally sexed, severely limit analyses of gender roles in past societies and may indeed bias them (Sofaer Derevenski 1997a). She asserts that "gender and gendered behaviours are age related, the cultural construction of 'social age' is cross-cut by gender ideology" (Sofaer Derevenski 1997a: 876). Thus the age-gender relationship is seen as a continuum, while biological sex is a dichotomous variable. Sofaer Derevenski (1997a: 877) writes that, "[i]n exploring gender in the archaeological record, we should look for continuous, rather than discrete variables, displaying different age-related changes in distribution between men and women. Conversely, in looking at sex we should identify discrete, rather than continuous variables, on the basis of presence and absence."

In her analysis of the site of Tiszapolgár-Basatanya, Sofaer Derevenski (1997a) first argues that, since there is a strong association between skeletal sex and burial on the right side (male) and on the left side (female), sidedness represents sex and can be used to determine the sex of juveniles. Then she explores associations for various artifacts. For example, during Period I ceramic vessels were interred with both males and females, although their patterned association with age differed. This, she argues, reflects a gendered pattern, which disappears during Period II, when there were no significant age-associations but an overall asymmetry between males and females. During Period II, "[d]ifferences between right- and left-side burials in their relationship with age disappear, as the continuous variable of gender transforms into discrete sex association; right-side burials now contain a significantly lower mean number of vessels than do left-side burials. . . . Thus the nature of gender relations changes over time as the role of ceramics changes from that of a mediator of gender and age relations, to a mediator of sex relations" (Sofaer Derevenski 1997a: 882). Her analysis was facilitated, of course, by the association between burial position and skeletal sex, which she reports as nearly perfect. For adults, one could simply base an age-related analysis upon associations between skeletal sex, age-at-death, and material culture or grave architecture, but the archaeological circumstances that facilitate age-gender-sex distinctions are relatively rare, especially in the precontact New World. This intricate analysis is clearly innovative in its approach to distinguishing sex and gender and in its pioneering effort in life-course analysis, as discussed below.

Another exemplary gendered analysis is that of Elizabeth Rega (2000), who studied skeletal attributes, grave goods, and spatial organization of the Early Bronze Age cemetery at Mokrin in Yugoslavia. It appears that individuals under the age of one were excluded from the cemetery. Grave goods seem heavily gendered, including those of adolescents. Interestingly, while girls as young as six years of developmental age were buried with weaving tools, males did not receive their gender-coded knives until they were adults, leading Rega (2000: 248) to conclude that "it may be that female children assume adult female activities far earlier than their male counterparts."

Contrasting with studies that derive gender from archaeological contexts, Sofaer (2006a: 113) contends that,

> [r]ather than inferring gender roles through grave goods, changes to the skeleton can be related to the ways that gender as a social institu-

tion impacts on the body. This produces bodies that are not more or less gendered, but rather bodies that are gendered in socially and temporally specific ways. Particular forms of gendered bodies are created through practice with the health consequences that this entails, as people express their social position and relation to others by way of their bodily *hexis* (personal manner and style of the body including deportment, gait, stance and gestures) (Bourdieu 1977), which itself becomes part of the body.

Sofaer cites the growing body of literature on activity-related changes in the skeleton, including musculoskeletal stress markers (MSMs), nonpathological joint changes, directional asymmetry, cross-sectional geometry, trauma, and pathological joint changes, as holding promise for identifying gendered activity differences. As Sofaer notes, a number of researchers stress that caution should be exercised in interpreting such attributes in terms of specific activities, as many of these "markers" have not been verified through clinical or natural experiments (Jurmain 1999; Knüsel et al. 1997; Rogers and Waldron 1995; Stirland 1991; see also Pearson and Buikstra 2006). The social biographies in White et al. (this volume) include several of the attributes discussed by Sofaer, notably MSMs and facet extensions, along with dental wear.

An impressive study of MSMs in the context of bioarchaeological gender studies is that of Jane Peterson (2002), who examines the sexual division of labor at the dawn of agriculture in the southern Levant. Her results indicate that labor was reorganized during the transition from hunting and gathering to mixed farming. She also argues for a more demanding lifestyle for both men and women at the point when domestic plants and animals entered the subsistence repertoire. Other exemplary analyses include an investigation of gendered activities and bone diaphyseal diameters by Patricia Bridges (1989) and an investigation of degenerative changes among the Inuit by Charles Merbs (1983). In addition, many authors, such as Christine Hastorf (1991), have studied gender and diet through bone chemistry. Hastorf's nuanced and detailed report is informed by ethnographic and ethnohistoric sources, as well as by numerous lines of archaeological evidence concerning chronological change in diet. In chapter 10, Torres-Rouff (this volume) develops a gendered perspective on cranial modification at the northern Chilean site of San Pedro de Atacama.

A focus upon *age* and identity is a relatively recent entry in scholarly discourse within both archaeology and bioarchaeology (Baxter 2005a,

2005b; Gilchrist 2000, 2004; Gowland 2006; Lucy 2005a; Sofaer Derevenski 1994, 1997a, 1997b, 2000). Certainly, age and aging have been studied previously, beginning in the late 1980s and 1990s, when second-wave feminists highlighted the role of aging adult women within families and communities (Gilchrist 2004). Children's roles also emerged as a subject of interest at this time, initially interpreted through the medium of material culture (Lillehammer 1989). The general study of age and identity, however, is more recent.

Age as a dimension of identity has been discussed to a limited extent in the preceding section on gender. There are three further important points, however, that deserve consideration. Chronologically, the first of these is the explicit interest in the archaeology of childhood (Baxter 2005a, 2005b; Kamp 2001; Sofaer Derevenski 1994, 2000) and the bioarchaeology of children (Lewis 2007). Stimulated by second-wave feminist theory, archaeological approaches to children and childhood have developed in relationship to gender studies and are explicitly designed to move beyond mortuary contexts (Baxter 2005a, 2005b; Gilchrist 2004). Nevertheless, as Megan Perry (2005) notes, analyses of children from cemeteries have tended to dominate archaeological studies of childhood overall.

In defining an archaeology of childhood, Jane Baxter (2005a: 95–96) is heavily influenced by Christopher Carr's (1995) cross-cultural study, which indicates that "age, particularly the distinction between child and adult, is one of two dominant social factors determining burial practices . . . grave location, body preparation, energy expenditure, and the number of burial types." Carr (1995) discovered that in the context of mortuary practice, age—primarily child/adult distinctions—was related not only to social categories but also to belief systems and worldview.

Further important influences on this developing field are the critiques of Sofaer Derevenski (1997b, 2000), who contends that children's burial treatment should not simply be assumed to reflect the parents' wishes and power. She argues for a more complex appreciation of children as social actors. Baxter (2005a) embraces this perspective and also promotes an archaeology of children that is not limited solely to mortuary contexts.

The Bioarchaeology of Children: Perspectives from Biological and Forensic Anthropology by Mary Lewis (2007) is the first such volume to be published expressly as bioarchaeology, although earlier anatomical treatments exist (Baker et al. 2005; Scheuer and Black 2000, 2004). While she emphasizes methods for analyzing children's remains and does not discuss funerary contexts, Lewis (2007) uses the methods that are required to enrich the bioarchaeological study of juvenile materials. A focus on the skeletons of children appears to have developed earlier in North America than in the United Kingdom. Lewis (2007: 3) cites a British interest in children from funerary contexts as occurring in the 1990s, which postdates similar developments in the United States by more than a decade (Cook 1979; Cook and Buikstra 1979).

A second, related trend has been the introduction of life-course analysis, which generally is defined as the cultural narrative of aging, including how societies experience, signify, and measure age. Age is not simply a biological process but also a cultural construction. The physiological changes that occur during growth, maturation, and senescence are understood culturally as a succession of life stages, characterized by different forms of knowledge, social roles, and symbolic meanings (Gilchrist 2000; Robb 2002). Multiple factors, such as gender and status, can influence the particular stages through which a person passes. In addition, as a person moves through life, various aspects of his or her identity wax and wane in importance. Thus, while age and aging are fundamental to life-course analysis, all other aspects of identity assume significance as they develop and transform over time.

Sofaer Derevenski's (1997a, 1997b) pioneering efforts to include children within archaeological studies have been discussed above. In these works, she introduces the concept of the "life-course" and contends that "[g]ender and age are intertwined throughout the life-course of the individual as gendered expectations, ideologies, self-perceptions and perceptions by others change, both from biological and social points of view" (Sofaer Derevenski 1997a: 876). More recently, in an introduction to a special issue of *World Archaeology* on human life-cycles, Gilchrist (2000: 327) emphasizes the scale and scope of such analyses and writes that "the papers in this volume adopt the lifecycle, life course or life history [approach], as a scale of analysis that is human in scope, and highlights the temporal and biographical measures of identity, memory and ageing."

While life-course analyses have only emerged within the past decade in archaeology, they have a longer history in the other social sciences. Gilchrist (2004), for example, traces life-course analyses to the sociological and gerontological concerns of the 1980s with aging and the aged in the West. Historians have tended to adopt a sociological perspective, recognizing that age, gender, ethnicity, and class are the four principal elements that structured individual and social experience in the past. Within anthropology, life-course analysis traces its heritage to notions of rites of passage, generally conceived as transitions between phases of the life cycle. Rather than viewing the human lifespan as a series of punctuated stages, since the 1980s anthropologists have placed increased emphasis upon aging as a continuum and the experience of aging (Gilchrist 2004).

The second and third waves of feminism have also influenced research on aging and the human life course. Second-wave feminism focused upon aging in adult women, while archaeology turned to household studies. From the third wave and a contemporary interest in the body emerged a concern for the development of differences, as they are embedded within the full life course (Gilchrist 2004).

Life-cycle applications in archaeology have been introduced by feminist scholars such as Lynn Meskell (1999), who argues that age-at-death was the principal structuring agent for the burials at the Egyptian New Kingdom site of Deir el Medina. In extending her analysis to gender, she argues for a life-cycle perspective, finding that transitions were more strongly marked for men than for women (Meskell 1999).

Terming his analytical procedure "osteobiography," John Robb (2002: 155) advocates biographical narratives that do not focus upon specific individuals but rather form "a cultural idea of what a human life should be. A first approximation of this is the succession of statuses an individual passes through during his or her lifespan. Each status usually has its own defining criteria, forms of knowledge, work, rights, duties, symbolic associations, and dangers." Robb (2002) states that his concept of osteobiography is loosely grounded in Frank Saul's (1972; see also Saul and Saul 1989) original use of the term. For Saul (1972), however, osteobiography meant an essentially forensic approach to data collection from individual skeletons, whereby numerous data categories relating to diverse topics such as diet, pathology, demography, and heritage were recorded for each skeleton during a single observational bout. Analytically these data classes were commonly merged to form a group perspective summed across single categories. Thus Saul's (1972) osteobiographies ended at approximately the same point that contemporary, explicitly population-oriented bioarchaeology (Buikstra 1977) did. In application, Robb's (2002) osteobiography more closely resembles Diane Hawkey's (1998) intensive analysis of the life of a single (diseased) individual. In other words, Saul's (1972) focus upon the individual was primarily methodological, while Robb's osteobiography (2002: 160) highlights "the cultural understanding of life events and . . . the history of human remains after death, a branch of study often relegated to specialized taphonomic studies." His case study uses a single adult female burial, Catignano I, to construct a narrative of her life, reflected against other knowledge of the Italian Neolithic. This approach is clearly aligned with life-course analyses that seek to identify normative experiences for individuals within a defined community or society. Ultimately, Robb (2002: 166) proceeds to define four "alternative biographical pathways" for the Italian Neolithic, which correspond to spatially and formally distinct interment procedures. While the assumption that these burial distinctions reflect different life histories rather than circumstances of death may seem problematic, Robb's approach is stimulating and creative.

In closing, Robb (2002: 168) evaluates the promise and prospective limitations of osteobiography:

> Burial and skeletal evidence provides an approach to the interaction of age and gender categories, to cultural modifications of the body, to the accumulation of relative seniority and the restriction of knowledge within a group, to important social processes such as illness, and to alternative life pathways. Using skeletons for this purpose requires us to develop paleopathological studies in new directions, particularly in investigating the experience of bodily processes such as illness and injury, in integrating skeletal data fully with archaeological data, and in paying more attention to when in the lifespan things such as trauma and cultural modification occurred.

While the notion that age is rarely considered in paleopathological analysis may seem odd to North American researchers, Robb's (2002) osteobiographic approach explicitly specifies the manner in which biographical narratives may be defined. As such, it resembles the life-course analyses recommended by other researchers.

While in 2000 Gilchrist was willing to accept a variety of terms—life cycle, life course, life history—for the ontogenetic approach she favors, by 2004 she advocated sole use of the term "life course." More specifically, Gilchrist (2004: 156) writes that "[t]he *life-course* model should be distinguished from the *life cycle*, which carries overtly biological and cross-cultural overtones. The contrast between life course and life cycle is greater than mere terminology. In the formative stages of theorizing age, this distinction may be as significant as that which contrasted the terms 'gender' and 'sex' in the fledgling discipline of gender archaeology."

The "life history" perspective, which emerged in evolutionary biology in the 1990s, seeks to investigate the human life course in evolutionary terms, emphasizing reproductive and survival strategies that emerge during ontogeny. In contrast to the life cycle and life history, the life course is historically situated and socially constructed—"a 'longitudinal' approach which examines trajectory and transition across the continuum of the human life, and which situates the human life span within social measures of time" (Gilchrist 2004: 156).

While not explicitly using the term "life course," White et al. (this volume) argue for a transformation of traditional osteobiography to social biography. Thus osteobiographical data on diet, health, activity, and long-distance mobility contribute more fully to studies of social identity.

An issue that should be raised in relationship to life-course studies is the frequent termination of "life" at the time of physical death. As noted above, Robb (2002) extends that period beyond death but links it to the taphonomic study of remains. Culturally, the life course of an individual may extend well after death. This concept is found in sociology as well (Hallam et al. 1999). In the West, various religious figures remain prominent within the lives of subsequent generations, some over millennia. Similarly, Ellen Bell (2002) contends that the founding couple of the Classic Copán dynasty was transformed after death into cosmic progenitors, symbolically linked to the sun (male) and moon (female). In other contexts, we are reminded that only a limited number of individuals become ancestors, who continue to be active participants in the social, political, economic, and religious affairs of the living (Buikstra and Charles 1999; Fortes 1965; Gluckman 1937; Morris 1991). Physical death leading to ancestorhood is only one more transition in the life course.

A third significant development in the bioarchaeological study of age, related to the life-course approach, is the deconstruction of age categories so that we do not impose our perceptions of important age transitions—either developmental or social—upon the past (Baxter 2005a; Gowland 2006). Gowland (2006: 143) explicitly distinguishes three "types" of age definitions that should not be confused:

1. Physiological/biological age (representing the physical ageing of the body).
2. Chronological age (corresponding to the amount of time that has passed from the moment of birth).
3. Social age (socially constructed norms concerning appropriate behavior and attitudes for an age group).

Gowland (2006: 144) also contrasts two ways to approach the study of age identity: (1) age grade/cohort and (2) life course. She prefers the latter, which she defines as the length of time from conception to death, unlike

Robb's (2002) extension after death. Next Gowland (2006: 145) develops a set of four important questions that bear attention in the study of age in the past:

- How did past populations conceptualize and structure their life course?
- What particular characteristics were each of these stages imbued with and how were they symbolised?
- How did age interact with other aspects of the social persona such as gender and status?
- How are these age transitions managed?

Gowland (2006) then addresses these questions by analyzing grave good assemblages by age group in early Anglo-Saxon cemeteries. In so doing, she establishes, for example, a shift in social status for females by the 13–17 year age group. More grave goods are interred with younger than with older women. Males appear to have attained adult status, as witnessed by the presence of swords in graves, by the age of 18. Gowland makes a further important point about age and the identity of the mourners. As an individual ages, it becomes more likely that she or he will be buried by offspring rather than by parents. Thus the observed decrease in grave wealth among older adults may reflect the relatively diminished economic resources of children and young adults.

The study of *disability* and identity is the least-developed research venue within either archaeological or bioarchaeological discourse. Early attempts to consider compassion and caring for infirm individuals, such as Ralph Solecki's (1971) discussion of Shanidar 1 and David Frayer et al.'s (1987) description of Romito 2's dwarfism, have been effectively critiqued by Katherine Dettwyler (1991) as anthropologically naïve. In recent years, scholars have attempted to develop more nuanced approaches.

Contemporary definitions of disability imply "a permanent or semi-permanent state that impinges on the ability to function 'normally' within society" (Vlahogiannis 1998: 15). Of course, "normal" encompasses a range of abilities within any society, and definitions may vary considerably across culturally distinct groups. In studies of disease or disability in relationship to identity, it is the reaction of *both* the individual and society to the condition that is of central significance.

In her study of disease in historical contexts, Isla Fay (2006) makes the point that there are two forms of disease. The first is personal and relates to physiological failure and is a subject for paleopathologists. The second is conceptual, consisting of language and discourse. This is the identity that an ill individual assumes based upon culture norms, and it may leave no archaeological residues. Fay also notes that there is a tension between paleo-pathologists, who apply biomedical knowledge to past cases, and historians of medicine, who address the invention of and social response to new diseases. She believes, however, that reconciliation seems possible (Fay 2006: 193).

> The fingers of historiography penetrate deeply into the cognitive plane of disease. But disease culture is not just about the sum of intellectual achievements, epistemic campaigns, or the diffusion of medical doctrine. It is also about praxis, manners, customs and *attitudes* at the level of individual interaction. Funerary archaeology has the potential to contribute significantly to our understanding of the "definition" and "experience" of bodily illness, as well as the "social response" to it: features that historians consider to be the quintessence of disease culture.

Morag Cross (1999) similarly contrasts a "social model of disability" (which parallels "disease culture" in Fay 2006) with the medical model, which is essentially the same as Fay's (2006) biomedical model. Cross also distinguishes between impairment, which is recorded in the skeletal record, and disability. She contends that "[i]n their concentration on 'fossilized disease' in the form of skeletal deformity, archaeologists dig up impairment, not disability. Joint disease is not a disability, it is an impairment. In archaeology there seems to be no concept of 'disability' according to the social model" (Cross 1999: 24). Thus, for Cross, the distinction between disability and impairment parallels that between gender and sex.

Defining disability identity in the past is a challenging task, especially in the absence of documentary evidence. Dettwyler (1991) argues, from the perspective of a medical anthropologist, that assuming compassion and caring when individuals with deforming pathology live into adulthood, for example, is unwarranted. Such interpretations reflect a (frequently unconscious) tendency to impose a "noble savage" image upon prehistoric peoples and also a prejudice against those who are infirm. From her ethnographic experience, Dettwyler (1991: 379–382) identifies the following five unwarranted assumptions:

1. The vast majority of a population's members are productive and self-sufficient most of the time.

2. Individuals who do not show skeletal/fossil evidence of impairments were not disabled.
3. A person with a physical impairment is, necessarily, nonproductive.
4. "Survival" of disabled individuals is indicative of "compassion."
5. Providing for, caring for, and facilitating the survival of a disabled individual is always the "compassionate" thing to do.

Dettwyler (1991) concludes, succinctly and pointedly, that we can learn practically nothing about compassion from the study of bones and artifacts.

In a more recent theme issue of the *Archaeological Review from Cambridge* (Finlay 1999a), based on a conference session designed to open a dialogue among archaeologists on the social construction of disability, Nyree Finlay (1999b: 3) optimistically states that "[t]here has to be more to an archaeological engagement with disability than simply the identification of impairment and its equation with discrimination." Although discrimination has replaced compassion, the assumed link is no more convincing. As Cross (1999) underscores, all the archaeological books she found in the library on ancient health and disease were shelved in the medical sections. The need for an archaeology of disability is great, but this is difficult to achieve for the many reasons cited by Dettwyler (1991) and discussed by Charlotte Roberts (1999).

The consensus from this conference session echoed Dettwyler's (1991) pessimistic opinion that little to nothing can be said about the individual or social construction of disability from the study of skeletal remains alone. A note of optimism emerged, however, when skeletons were examined within their archaeological contexts. Especially convincing is Knüsel's (1999) "test" of Dettwyler's hypothesis (1991: 384) that "[t]he palaeopathological analysis of skeletal remains can tell us about the physical impairment, from which we may be able to infer the extent of an individual's disability. Whether or not an individual was 'handicapped' by his disability cannot be determined from archaeological evidence alone." In this framework, "handicapped" references the social construction of disability.

Knüsel (1999) examines the remains of three medieval English skeletons, all of which presented deforming pathology that would have affected mobility. All three individuals were adult males. One had suffered traumatic injury to the right knee, which had rotated the leg (tibia) externally. The second fully adult male had a slipped proximal right femur epiphysis, a condition that usually occurs in adolescence. Finally, an elderly male with leprosy also possessed a series of healed fractures, including the right femur, which resulted in limb asymmetries. The first two men had been interred in privileged situations; the second, in fact, was buried beneath the altar of St. Giles church with a mortuary chalice and paten. Knüsel (1999) suggests that the deformity may have advantaged this priest by linking him to his patron, St. Giles, who had also suffered a deforming injury. The leper, by contrast, was segregated with others so afflicted within the cemetery of a *leprosarium*. In terms of a socially constructed disability persona, for this individual, the infectious disease assumed primary significance in the eyes of others.

Thus the study of remains within their funerary contexts, including grave location as a significant dimension of variability, permits finer—if not perfect—definition of disability within social discourse, a conclusion also reached by Roberts (1999) and Theya Molleson (1999). Molleson (1999: 69) summarizes the archaeological study of disability by arguing that "[a]n understanding of attitudes to disability in the archaeological context can really only be inferred from the treatment of the dead. Few disabilities actually affect the skeleton so that our insight is necessarily limited." Consequently, through the study of human remains *and* their archaeological contexts, we may be able to address societal definitions of disability for those individuals who register infirmities skeletally. Determinations of self-definitions are more difficult. Because social treatment inevitably affects self-perception, however, one may assume that the St. Giles priest did *not* think of himself as disabled, while the anonymous leper accepted society's pronouncement, though he may have had little choice in the matter.

Hawkey (1998), in an exquisitely detailed study of a diseased individual (GQ 391) who died during young-middle adulthood and was interred at Gran Quivira Pueblo (New Mexico) during the Late Period (AD 1560–1672), illustrates the type of osteobiographic approach to health and life course called for by Robb (2002). By combining evidence of clinical disease progression, carefully diagnosed as a form of systemic juvenile rheumatoid arthritis, with joint mobility and musculoskeletal stress markers (MSMs), Hawkey (1998) infers that GQ 391 increasingly lost mobility with age during childhood and ultimately had to be cared for during the final portion of his life, beginning in young adulthood. Impairment is clearly evident in osseous deformities that limited motion. Hawkey argues that she can estimate disability by comparing the MSMs of GQ 391 to other individuals, though this clinical perspective obviously does not speak to societal attitudes. Eschewing an inference of compassion, she reasonably concludes that at least one individual had to care for GQ 391 for a long period. As Hawkey notes, aside from burial location, all aspects of corpse treatment (such as body position, orientation, and grave goods) were consistent with those of other adults. Only the location of his interment in a place normally reserved for children hints at special social status for GQ 391 as a disabled individual. This further underscores the significance of burial context in meeting the challenge of inferring the identity of (socially) disabled people in the absence of historical records.

The Language of Identity

In conclusion, we would like to present definitions of the following five terms as they are used in the archaeological and bioarchaeological literature on identity: agency, embodiment, materiality, personhood, and selfhood. For those newly entering the literature on identity from a background in physical anthropology, such terminology may appear mysterious. We concentrate on identifying key references and providing a definition thus derived. In some cases, however, as in agency and embodiment, there is no consensus definition. The best advice for scholars tempted to use such terms is to define meaning explicitly at the outset of any publication.

Agency is multiscalar and can be attributed to individuals (Hodder 2000) or to communities (Wobst 2000), thus being relevant to both part I and part II of this volume. The term is widely used across a wide range of theoretical perspectives, from processualist to postprocessualist. It can be defined broadly or more narrowly, but there is little agreement among archaeologists. Definitions vary, from those that emphasize intentionality to others that stress agency's nondiscursive manner (Dobres and Robb 2000b). The influential Marcia-Anne Dobres and John Robb (2000a) volume, for example, was expressly designed to start conversations about the nature of agency. While most would concur that agency is a quality of action, rather than the action itself, beyond that there is little agreement.

A more recent treatment of agency includes papers that focus upon its recursive relationship with structure (Hegmon and Kulow 2005; Joyce and Lopiparo 2005). Such studies follow Anthony Giddens' (1984) concept of structuration. As Michelle Hegmon and Stephanie Kulow (2005: 316) observe, "structure exists only in so far as it is reproduced by the conduct of actors, so the exercise of agency can reproduce, reinforce, or transform structure." On the other hand, Rosemary Joyce and Jeanne Lopiparo (2005) argue against the notion that agency and structure alternate. Obviously, consensus has yet to be found.

Embodiment concerns the manner in which the body is shaped individually and socially during ontogeny (Joyce 2000; Meskell 1999). This concept has developed in the context of feminist theory. It explicitly opposes the Cartesian mind/body duality and instead considers the body as a product of lived experience. Embodiment as a concept is frequently traced by practitioners to the essay by Marcel Mauss (1979 [1950]: 97) on "body techniques," which he defined in terms of "the ways in which from society to society men know how to use their bodies." The work of Mary Douglas (1966) is also frequently cited, in its emphasis upon the body as a metaphor for society (Meskell 1998).

As with several other terms related to identity, there is a nuanced dispute over the proper definition of embodiment and the extent of its usefulness (Ingold 1998; Meskell 1998; Montserrat 1998). Tim Ingold (1998), for example, points out that usage frequently emphasizes social construction without also referencing the organic nature of the biological individual. He therefore prefers the term "organism," which could be argued to be overly biological.

Lynn Meskell (1998) and Dominic Montserrat (1998) criticize archaeology's reliance on Foucauldian interpretations, which they say unduly privileges social power over the body and thus diminishes the role of individuality and intentionality. Meskell (1998: 141) declares as follows:

Archaeology has been seduced by Foucauldian notions of control, where power relations are mapped on the body as a surface which can be analysed as a forum for display. Despite the eager and enthusiastic adoption in archaeology of these ideas, replete with post-modern posturing and sanctioned by feminist practitioners, there is still the implicit adoption of binary, dichotomous and essentially Cartesian notions of rigid sex typing. Such a critique crystallised through specific readings of masculinist theory and its critique of the social construction of hegemonic masculinity through the ascendancy of the mind over bodily or emotional experience. From these readings, it could be argued that the current preoccupation with control and elaboration is a typically androcentric, externalized separation of mind, body and emotion.

Meskell (1998: 148) also emphasizes that the body is "not merely constrained by or invested with social relations, but also forms a basis for and contributes towards these relations." Meskell (1998: 159) further argues that "[a]n embodied body represents, and is, a lived experience where the interplay of irreducible natural, social, cultural and psychical phenomena are

brought to fruition through each individual's resolution of external structures, embodied experience and choice." However archaeologically challenging it may be to access without the presence of textual referents, this
synergistic perspective is favored here.

Initially adopted by archaeologists, embodiment has recently become
visible in bioarchaeological discourse. Sofaer (2006a) links the concept of
embodiment to that of *materiality*. In *The Body as Material Culture*, Sofaer
(2006a: 66) argues for materiality "as the foundation of embodiment where
the effects of the materiality of specific bodies lead to particular embodied
experiences of individuals." She contends that treating the body as a biological entity artificially separates human remains from their archaeological contexts. She prefers instead to treat the body as material culture and
emphasizes plasticity as a key attribute. This plasticity permits the registra-
tion of activity and disease, for example, as discussed above. Asserting that
"[b]odies are *material* and *social*, just as objects are," Sofaer (2006a: 85)
maintains that "[f]rom a methodological perspective, bodies and objects
fall into the same archaeological domain." Thus the boundary between persons and artifacts is breached, and the tension between archaeologists and
osteoarchaeologists disappears.

Grounded in postmodern theories of the body, this approach to bioar-
chaeology/osteoarchaeology invites concern for several reasons. First, it
creates an artificial direct link between the body and the grave, even though
the latter is a creation of the mourners, not the deceased. Further, it en-
courages researchers to ignore the other biological processes that shape the
human body and the form that remains take. Finally, it limits bioarchaeological inquiry to the effects of plasticity, when there are many other classes
of information that may be derived from osteological analysis, not the least
of which involves heritage, an important factor in the study of identity as
the papers in part I of this volume by Stojanowski, Nystrom, and Sutter
demonstrate.

Embodiment is the subject of chapters by White et al. (this volume) and
Duncan (this volume). Duncan makes the novel argument that cranial
deformation among the Maya, commonly considered the embodiment of
social status, kin relations, or some other form of group identity, reflects
an effort to prevent soul loss during childhood rites of passage (the *héetz-
méek'* ceremony). In this case, ethnographic sources deepen appreciation
of the interplay of social and biological forces in embodiment. In chapter
7, White et al. (this volume) directly engage embodiment's relationship to
bioarchaeology, thus enriching their social biography of three individuals

buried at the Maya site of Lamanai. White et al. follow a more traditional interpretation of cranial deformation as a statement of group identity, as do Torres-Rouff (this volume) and Knudson and Blom (this volume).

In a classic work that distinguishes between the individual, self, and person, Grace Harris (1989) defines person and *personhood* in terms of the possession of agency and social recognition. The self, by contrast, is the locus of experience, "psychologistic" in Harris' terms. Last, an individual is a single human being, "biologistic." Her definition of personhood extends to deceased individuals treated as "agents-in-society" (Harris 1989: 602). Ancestors, deities, nonhuman animals, and objects that we would consider inanimate may be persons if their actions affect human lives.

In *The Archaeology of Personhood: An Anthropological Approach*, Chris Fowler (2004) follows Harris' definitional breadth. He explicitly defines person and personhood in the following manner (Fowler 2004: 7):

> *Person* is used to refer to any entity, human or otherwise, which may be conceptualized and treated as a person. A person is frequently composed through the temporary association of different aspects. These aspects may include features like mind, spirit or soul as well as a physical body, and denote the entity as having a form of agency. Exactly who or what may or may not be a person is contextually variable.
>
> *Personhood* in its broadest definition refers to the condition or state of being a person, as it is understood in any specific context. Persons are constituted, de-constituted, maintained and altered in social practices through life and after death. This process can be described as the ongoing attainment of personhood. Personhood is frequently understood as a condition that involves constant change, and key transformations to the person occur throughout life and death. People may pass from one state or stage of personhood to another. Personhood is attained and maintained through relationships not only with other human beings but with things, places, animals and the spiritual features of the cosmos. Some of these may also emerge as persons through this engagement. People's own social interpretations of personhood and of the social practices through which personhood is realized shape their interactions in a reflexive way, but personhood remains a mutually constituted condition.

Two key points that emerge from Harris' (1989) and Fowler's (2004) definitions are the role of agency and the transformative nature of personhood. Persons are perceived to have agency and they pass from one state to another, as in the life-course analytical model discussed above. Gilchrist (2004) connects this focus upon personhood to the emergence of theories of the body in the late 1980s and 1990s.

While Fowler's definition is very specific, not all scholars writing about identity embrace it in its breadth. The term "personhood" is frequently invoked without definition, more narrowly construed as a human person than in the context defined by Harris (1989) and Fowler (2004).

Selfhood, as an extension of Harris' (1989: 601) definition of self, references the individual as "a locus of experience, including experience of that human's own someoneness." This concept has been adapted to archaeological contexts by Meskell (1998: 155) among others as a term referencing individuals or families. In this formulation, Meskell (1998: 155) moves away from defining or prioritizing dimensions of identity, such as "age, status, class, gender, ethnicity or marital status," and argues that by emphasizing selfhood, "single burials of children, couples, family groups . . . may show how notions of identity or constructions of self were embodied." Her example of deformed children with elaborate offerings from the poorest cemetery at Deir el Medina, where written histories exist for some individuals, forms a compelling case of bereavement and commemoration (Meskell 1998). Sarah Tarlow's (1999) research on the Orkney Islands during the eighteenth and nineteenth centuries AD offers another example of selfhood and individual identities in the context of material culture, mortuary patterns, and life histories of named persons. The application of this concept would appear most challenging, but not impossible, in the absence of historical documents.

In this chapter, we have provided an introduction to bioarchaeological research on identity, focusing on the dimensions of identity commonly investigated and the terms employed. While cultural anthropologists and archaeologists have long studied religion, social status, and ethnicity, interest in gender, age, and disability developed only after the injection of feminist theory into archaeology in the 1980s. The recognition that these various dimensions are intertwined has resulted in the current emphasis on identity, as referencing the multiple forms of social affiliation that define individuals and communities. The ultimate goal of identity research is thus to incorporate "all those vectors of difference by which individuals are named and subjectified" (Meskell 2001: 188). Although bioarchaeology has only recently entered this discourse, it has the potential to contribute significantly to the investigation of identity. As the case studies in this volume illustrate, the most effective analyses of identity examine a variety of archaeological and skeletal data, consider specific cultural and historical contexts, and draw on social theory.

Acknowledgments

Considerable basic research for this chapter occurred while the senior author was at Durham University (UK) during early 2007, on a fellowship sponsored by the Institute of Advanced Study (IAS). Pleasant and productive conversations with other fellows at Cosin's Hall, where the IAS is housed, as well as with various members of the Department of Archaeology at Durham University, stimulated a much more in-depth treatment of identity and bioarchaeology than was initially envisioned. Those who made special contributions to this project include Professors Rebecca Gowland, Andreas-Holger Maehle, and Charlotte Roberts from Durham University. While productive, this extended research required multiple deadline extensions, and the authors would therefore like to express their appreciation to the editors both for the invitation to participate in this excellent volume and for the gentleness of their reminders.

References Cited

Arnold, Bettina, and Nancy L. Wicker, editors. 2001. *Gender and the Archaeology of Death*. Walnut Creek, Calif.: AltaMira Press.

Babić, Staša. 2005. "Status Identity and Archaeology." In *The Archaeology of Identity: Approaches to Gender, Age, Status, Ethnicity, and Religion*, edited by M. Díaz-Andreu, S. Lucy, S. Babić, and D. N. Edwards, 67–85. London: Routledge.

Baker, Brenda J., Tosha L. Dupras, and Matthew W. Tocheri. 2005. *The Osteology of Infants and Children*. College Station: Texas A&M University Press.

Barrett, James H., Roelf P. Beukens, and Rebecca A. Nicholson. 2001. "Diet and Ethnicity during the Viking Colonization of Northern Scotland: Evidence from Fish Bones and Stable Carbon Isotopes." *Antiquity* 75: 145–154.

Barth, Fredrik. 1969. "Introduction." In *Ethnic Groups and Boundaries: The Social Organization of Culture Difference*, edited by F. Barth, 9–38. Boston: Little, Brown and Company.

Baxter, Jane E. 2005a. *The Archaeology of Childhood: Children, Gender, and Material Culture*. Walnut Creek, Calif.: AltaMira Press.

———, editor. 2005b. *Children in Action: Perspectives on the Archaeology of Childhood*. Archeological Papers of the American Anthropological Association 15. Washington, D.C.: American Anthropological Association.

Beck, Lane A., editor. 1995. *Regional Approaches to Mortuary Analysis*. New York: Plenum Press.

Bell, Ellen E. 2002. "Engendering a Dynasty: A Royal Woman in the Margarita Tomb, Copan." In *Ancient Maya Women*, edited by T. Arden, 89–104. Walnut Creek, Calif.: AltaMira Press.

Bloch, Maurice. 1992. *Prey into Hunter: The Politics of Religious Experience*. Cambridge: Cambridge University Press.

Bond, Julie M., and Fay L. Worley. 2006. "Companions in Death: The Roles of Animals in Anglo-Saxon and Viking Cremation Rituals in Britain." In *Social Archaeology of Funerary Remains*, edited by R. Gowland and C. Knüsel, 89–98. Oxford: Oxbow Books.

Bourdieu, Pierre. 1977. *Outline of a Theory of Practice*. Cambridge: Cambridge University Press.

Bridges, Patricia S. 1989. "Changes in Activities with the Shift to Agriculture in the Southeastern United States." *Current Anthropology* 30: 385–394.

Brown, James A., editor. 1971. *Approaches to the Social Dimensions of Mortuary Practices*. Memoirs of the Society for American Archaeology 25. Washington, D.C.: Society for American Archaeology.

Buikstra, Jane E. 1977. "Biocultural Dimensions of Archaeological Study: A Regional Perspective." In *Biocultural Adaptation in Prehistoric America*, edited by R. L. Blakely, 67–84. Athens: University of Georgia Press.

———. 1991. "Out of the Appendix and into the Dirt: Comments on Thirteen Years of Bioarcheological Research." In *What Mean These Bones?: Studies in Southeastern Bioarchaeology*, edited by M. L. Powell, P. S. Bridges, and A. M. Mires, 172–188. Tuscaloosa: University of Alabama Press.

Buikstra, Jane E., and Douglas K. Charles. 1999. "Centering the Ancestors: Cemeteries, Mounds, and Sacred Landscapes of the Ancient North American Midcontinent." In *Archaeologies of Landscape*, edited by W. Ashmore and A. B. Knapp, 201–228. Oxford: Blackwell.

Butler, Judith. 1990. *Gender Trouble: Feminism and the Subversion of Identity*. New York: Routledge.

Carr, Christopher. 1995. "Mortuary Practices: Their Social, Philosophical-Religious, Circumstantial, and Physical Determinants." *Journal of Archaeological Method and Theory* 2: 105–200.

Chapman, Robert, Ian Kinnes, and Klaus Randsborg, editors. 1981. *The Archaeology of Death*. Cambridge: Cambridge University Press.

Cheek, Charles D., and Amy Friedlander. 1990. "Pottery and Pig's Feet: Space, Ethnicity, and Neighborhood in Washington, D.C., 1880–1940." *Historical Archaeology* 24: 34–60.

Conkey, Margaret W., and Janet D. Spector. 1984. "Archaeology and the Study of Gender." *Advances in Archaeological Method and Theory* 7: 1–38.

Cook, Della C. 1979. "Subsistence Base and Health in Prehistoric Illinois Valley: Evidence from the Human Skeleton." *Medical Anthropology* 3: 109–124.

Cook, Della C., and Jane E. Buikstra. 1979. "Health and Differential Survival in Pre-

Historic Populations: Prenatal Dental Defects." *American Journal of Physical Anthropology* 51: 649–664.

Crabtree, Pam J. 1990. "Zooarchaeology and Complex Societies: Some Uses of Faunal Analysis for the Study of Trade, Social Status, and Ethnicity." In *Archaeological Method and Theory*, vol. 2, edited by M. B. Schiffer, 155–205. Tucson: University of Arizona Press.

Cross, Morag. 1999. "Accessing the Inaccessible: Disability and Archaeology." Theme issue, "Disability and Archaeology," *Archaeological Review from Cambridge* 15: 7–30.

Dettwyler, Katherine A. 1991. "Can Paleopathology Provide Evidence for 'Compassion'?" *American Journal of Physical Anthropology* 84: 375–384.

Díaz-Andreu, Margarita. 2005. "Gender Identity." In *The Archaeology of Identity: Approaches to Gender, Age, Status, Ethnicity, and Religion*, edited by M. Díaz-Andreu, S. Lucy, S. Babić, and D. N. Edwards, 13–42. London: Routledge.

Díaz-Andreu, Margarita, and Sam Lucy. 2005. "Introduction." In *The Archaeology of Identity: Approaches to Gender, Age, Status, Ethnicity, and Religion*, edited by M. Díaz-Andreu, S. Lucy, S. Babić, and D. N. Edwards, 1–12. London: Routledge.

Dobres, Marcia-Anne, and John Robb, editors. 2000a. *Agency in Archaeology*. London: Routledge.

———. 2000b. "Agency in Archaeology: Paradigm or Platitude?" In *Agency in Archaeology*, edited by M.-A. Dobres and J. Robb, 3–17. London: Routledge.

Douglas, Mary. 1966. *Purity and Danger: An Analysis of Concepts of Pollution and Taboo*. London: Routledge and Kegan Paul.

Durkheim, Emile. 1915. *The Elementary Forms of Religious Life*. Translated by J. W. Swain. London: Allen and Unwin.

Eliade, Mircea. 1987 [1957]. *The Sacred and Profane: The Nature of Religion*. San Diego: Harcourt Brace and Company.

Evans-Pritchard, E. E. 1974 [1956]. *Nuer Religion*. Oxford: Oxford University Press.

Fay, Isla. 2006. "Text, Space and the Evidence of Human Remains in English Late Medieval and Tudor Disease Culture: Some Problems and Possibilities." In *Social Archaeology of Funerary Remains*, edited by R. Gowland and C. Knüsel, 190–208. Oxford: Oxbow Books.

Finlay, Nyree, editor. 1999a. Theme issue, "Disability and Archaeology," *Archaeological Review from Cambridge* 15.

———. 1999b. "Disabling Archaeology: An Introduction." Theme issue, "Disability and Archaeology," *Archaeological Review from Cambridge* 15: 1–6.

Fortes, Meyer. 1965. "Some Reflections on Ancestor Worship in Africa." In *African Systems of Thought*, edited by M. Fortes and G. Dieterlen, 122–144. Oxford: Oxford University Press.

Fowler, Chris. 2004. *The Archaeology of Personhood: An Anthropological Approach*. London: Routledge.

Frayer, David W., William A. Horton, Roberto Macchiarelli, and Margherita Mussi. 1987. "Dwarfism in an Adolescent from the Italian Late Upper Paleolithic." *Nature* 330: 60–62.

Frazier, James G. 1890. *The Golden Bough*. London: Macmillan.

Geertz, Clifford. 1973. *The Interpretation of Cultures: Selected Essays by Clifford Geertz.* New York: Basic Books.

Gero, Joan M., and Margaret W. Conkey, editors. 1991. *Engendering Archaeology: Women and Prehistory.* Oxford: Basil Blackwell.

Giddens, Anthony. 1984. *The Constitution of Society: Outline of the Theory of Structuration.* Berkeley: University of California Press.

Gilchrist, Roberta. 1994. *Gender and Material Culture: The Archaeology of Religious Women.* London: Routledge.

———. 1999. *Gender and Archaeology: Contesting the Past.* London: Routledge.

———. 2000. "Archaeological Biographies: Realizing Human Lifecycles, -Courses, and -Histories." Theme issue, "Human Lifecycles," *World Archaeology* 31: 325–328.

———. 2004. "Archaeology and the Life Course: A Time and Age for Gender." In *A Companion to Social Archaeology*, edited by L. Meskell and R. W. Preucel, 142–160. Oxford: Blackwell.

Gluckman, Max. 1937. "Mortuary Customs and the Belief in Survival after Death among the South-eastern Bantu." *Bantu Studies* 11: 117–136.

Goldstein, Lynne. 2006. "Mortuary Analysis and Bioarchaeology." In *Bioarchaeology: The Contextual Analysis of Human Remains*, edited by J. E. Buikstra and L. A. Beck, 375–387. New York: Academic Press.

Gowland, Rebecca. 2006. "Ageing the Past: Examining Age Identity from Funerary Evidence." In *Social Archaeology of Funerary Remains*, edited by R. Gowland and C. Knüsel, 143–154. Oxford: Oxbow Books.

Gowland, Rebecca, and Christopher Knüsel, editors. 2006a. *Social Archaeology of Funerary Remains.* Oxford: Oxbow Books.

———. 2006b. "Introduction." In *Social Archaeology of Funerary Remains*, edited by R. Gowland and C. Knüsel, ix–xiv. Oxford: Oxbow Books.

Hallam, Elizabeth, Jenny Hockey, and Glennys Howarth. 1999. *Beyond the Body: Death and Social Identity.* London: Routledge.

Harris, Grace G. 1989. "Concepts of Individual, Self, and Person in Description and Analysis." *American Anthropologist* 91: 599–612.

Hastorf, Christine A. 1991. "Gender, Space and Food in Prehistory." In *Engendering Archaeology: Women and Prehistory*, edited by J. M. Gero and M. W. Conkey, 132–159. Oxford: Blackwell.

Hawkey, Diane E. 1998. "Disability, Compassion, and the Skeletal Record: Using Musculoskeletal Stress Markers (MSM) to Construct an Osteobiography from Early New Mexico." *International Journal of Osteoarchaeology* 8: 326–340.

Hegmon, Michelle. 2003. "Setting Theoretical Egos Aside: Issues and Theory in North American Archaeology." *American Antiquity* 68: 213–243.

Hegmon, Michelle, and Stephanie Kulow. 2005. "Painting as Agency, Style as Structure: Innovations in Mimbres Pottery Designs from Southwest New Mexico." *Journal of Archaeological Method and Theory* 12: 313–334.

Hodder, Ian. 1978. "Simple Correlations between Material Culture and Society: A Review." In *The Spatial Organization of Culture*, edited by I. Hodder, 3–24. London: Duckworth.

———. 1982. *Symbols in Action.* Cambridge: Cambridge University Press.

———. 1984. "Burials, Houses, Women, and Men in the European Neolithic." In *Ideology, Power, and Prehistory,* edited by D. Miller and C. Tilley, 51–68. Cambridge: Cambridge University Press.

———. 2000. "Agency and Individuals in Long-term Processes." In *Agency in Archaeology,* edited by M.-A. Dobres and J. Robb, 21–33. London: Routledge.

Ingold, Tim. 1998. "From Complementarity to Obviation: On Dissolving the Boundaries between Social and Biological Anthropology, Archaeology and Psychology." *Zeitschrift für Ethnologie* 123: 21–52.

Insoll, Timothy, editor. 2001. *Archaeology and World Religion.* London: Routledge.

———. 2004. *Archaeology, Ritual, Religion.* London: Routledge.

———. 2007. "Introduction: Configuring Identities in Archaeology." In *The Archaeology of Identities: A Reader,* edited by T. Insoll, 1–18. London: Routledge.

Jones, Siân. 1997. *The Archaeology of Ethnicity: Constructing Identities in the Past and Present.* London: Routledge.

———. 2007. "Discourses of Identity in the Interpretation of the Past." In *The Archaeology of Identities: A Reader,* edited by T. Insoll, 44–58. London: Routledge. [Originally published in *Cultural Identity and Archaeology,* edited by P. Graves-Brown, S. Jones, and C. Gamble, 62–80. London: Routledge, 1996.]

Joyce, Rosemary A. 2000. "Girling the Girl and Boying the Boy: The Production of Adulthood in Ancient Mesoamerica." *World Archaeology* 31: 473–483.

Joyce, Rosemary A., and Jeanne Lopiparo. 2005. "Postscript: Doing Agency in Archaeology." *Journal of Archaeological Method and Theory* 12: 365–374.

Jurmain, Robert. 1999. *Stories from the Skeleton: Behavioral Reconstruction in Human Osteology.* Amsterdam: Gordon and Breach.

Kamp, Kathryn A. 2001. "Where Have All the Children Gone?: The Archaeology of Childhood." *Journal of Archaeological Method and Theory* 8: 1–34.

Knüsel, Christopher J. 1999. "Orthopaedic Disability: Some Hard Evidence." Theme issue, "Disability and Archaeology," *Archaeological Review from Cambridge* 15: 31–53.

Knüsel, Christopher J., Sonia Göggel, and David Lucy. 1997. "Comparative Degenerative Joint Disease of the Vertebral Column in the Medieval Monastic Cemetery of the Gilbertine Priory of St. Andrew, Fishergate, York, England." *American Journal of Physical Anthropology* 103: 481–495.

Kroeber, Alfred L. 1927. "Disposal of the Dead." *American Anthropologist* 29: 308–315.

Lewis, Mary E. 2007. *The Bioarchaeology of Children: Perspectives from Biological and Forensic Anthropology.* Cambridge: Cambridge University Press.

Lillehammer, Grete. 1989. "A Child Is Born: The Child's World in an Archaeological Perspective." *Norwegian Archaeological Review* 22: 89–105.

Lucy, Sam. 2005a. "The Archaeology of Age." In *The Archaeology of Identity: Approaches to Gender, Age, Status, Ethnicity, and Religion,* edited by M. Díaz-Andreu, S. Lucy, S. Babić, and D. N. Edwards, 43–66. London: Routledge.

———. 2005b. "Ethnic and Cultural Identities." In *The Archaeology of Identity: Approaches to Gender, Age, Status, Ethnicity, and Religion,* edited by M. Díaz-Andreu, S. Lucy, S. Babić, and D. N. Edwards, 86–109. London: Routledge.

Malinowski, Bronislaw. 1954. "Magic, Science and Religion." In *Magic, Science and Religion and Other Essays*, 17–87. Garden City, N.Y.: Doubleday.

Mauss, Marcel. 1979 [1950]. "The Notion of Body Techniques." In *Sociology and Psychology: Essays*, translated by B. Brewster, 97–105. London: Routledge and Kegan Paul.

Meiklejohn, Chris, Erik B. Petersen, and Verner Alexandersen. 2000. "The Anthropology and Archaeology of Mesolithic Gender in the Western Baltic." In *Gender and Material Culture in Archaeological Perspective*, edited by M. Donald and L. Hurcombe, 222–237. New York: Palgrave Macmillan.

Merbs, Charles F. 1983. *Patterns of Activity-Induced Pathology in a Canadian Inuit Population*. Archaeological Survey of Canada Paper. Mercury Series, No. 119. Ottawa: National Museums of Canada.

Meskell, Lynn. 1998. "The Irresistible Body and the Seduction of Archaeology." In *Changing Bodies, Changing Meanings: Studies on the Human Body in Antiquity*, edited by D. Montserrat, 139–161. London: Routledge.

———. 1999. *Archaeologies of Social Life: Age, Sex, Class, Et Cetera in Ancient Egypt*. Oxford: Blackwell.

———. 2001. "Archaeologies of Identity." In *Archaeological Theory Today*, edited by I. Hodder, 187–213. Cambridge: Polity.

Miller, Daniel, and Christopher Tilley, editors. 1984. *Ideology, Power, and Prehistory*. Cambridge: Cambridge University Press.

Molleson, Theya. 1999. "Archaeological Evidence for Attitudes to Disability in the Past." Theme issue, "Disability and Archaeology," *Archaeological Review from Cambridge* 15: 69–77.

Montgomery, Janet, and Jane A. Evans. 2006. "Immigrants on the Isle of Lewis—Combining Traditional Funerary and Modern Isotope Evidence to Investigate Social Differentiation, Migration and Dietary Change in the Outer Hebrides of Scotland." In *Social Archaeology of Funerary Remains*, edited by R. Gowland and C. Knüsel, 122–142. Oxford: Oxbow Books.

Montserrat, Dominic. 1998. "Introduction." In *Changing Bodies, Changing Meanings: Studies on the Human Body in Antiquity*, edited by D. Montserrat, 1–9. London: Routledge.

Morris, Ian. 1991. "The Archaeology of Ancestors: The Saxe/Goldstein Hypothesis Revisited." *Cambridge Archaeological Journal* 1: 147–169.

Parker Pearson, Michael. 1982. "Mortuary Practices, Society, and Ideology: An Ethnoarchaeological Study." In *Symbolic and Structural Archaeology*, edited by I. Hodder, 99–113. Cambridge: Cambridge University Press.

———. 2000. *The Archaeology of Death and Burial*. College Station: Texas A&M University Press.

Pearson, Osbjorn M., and Jane E. Buikstra. 2006. "Behavior and the Bones." In *Bioarchaeology: The Contextual Analysis of Human Remains*, edited by J. E. Buikstra and L. A. Beck, 207–225. New York: Academic Press.

Perry, Megan A. 2005. "Redefining Childhood through Bioarchaeology: Toward an Archaeological and Biological Understanding of Children in Antiquity." In *Children in Action: Perspectives on the Archaeology of Childhood*, edited by J. E. Baxter, 89–111.

Archeological Papers of the American Anthropological Association 15. Washington, D.C.: American Anthropological Association.

Peterson, Jane. 2002. *Sexual Revolutions: Gender and Labor at the Dawn of Agriculture.* Walnut Creek, Calif.: AltaMira Press.

Rakita, Gordon F. M., and Jane E. Buikstra. 2008. *An Archaeological Perspective on Ritual, Religion, and Ideology from American Antiquity and Latin American Antiquity.* Society for American Archaeology Reader Series. Washington, D.C.: Society for American Archaeology.

Rakita, Gordon F. M., Jane E. Buikstra, Lane A. Beck, and Sloan Williams, editors. 2005. *Interacting with the Dead: Perspectives on Mortuary Archaeology for the New Millennium.* Gainesville: University Press of Florida.

Rappaport, Roy A. 1968. *Pigs for the Ancestors.* New Haven: Yale University Press.

———. 1999. *Ritual and Religion in the Making of Humanity.* Cambridge: Cambridge University Press.

Rega, Elizabeth. 2000. "The Gendering of Children in the Early Bronze Age Cemetery at Mokrin." In *Gender and Material Culture in Archaeological Perspective,* edited by M. Donald and L. Hurcombe, 238–249. New York: Palgrave Macmillan.

Robb, John. 2002. "Time and Biography: Osteobiography of the Italian Neolithic Lifespan." In *Thinking through the Body: Archaeologies of Corporeality,* edited by Y. Hamilakis, M. Pluciennik, and S. Tarlow, 153–171. London: Kluwer Academic/Plenum.

Roberts, Charlotte. 1999. "Disability in the Skeletal Record: Assumptions, Problems, and Some Examples." Theme issue, "Disability and Archaeology," *Archaeological Review from Cambridge* 15: 79–97.

Rogers, Juliet, and Tony Waldron. 1995. *A Field Guide to Joint Disease in Archaeology.* Chichester: Wiley.

Saul, Frank P. 1972. *The Human Skeletal Remains from Altar de Sacrificios: An Osteobiographic Analysis.* Papers of the Peabody Museum of Archaeology and Ethnology 63(2). Cambridge, Mass.: Peabody Museum of Archaeology and Ethnology, Harvard University.

Saul, Frank P., and Julie M. Saul. 1989. "Osteobiography: A Maya Example." In *Reconstruction of Life from the Skeleton,* edited by Mehmet Y. İşcan and Kenneth A. R. Kennedy, 287–302. New York: Alan R. Liss.

Scheuer, Louise, and Sue Black. 2000. *Developmental Juvenile Osteology.* London: Academic Press.

———. 2004. *The Juvenile Skeleton.* London: Academic Press.

Schmidt, Robert A., and Barbara L. Voss, editors. 2000. *Archaeologies of Sexuality.* London: Routledge.

Scott, Elizabeth M. 2001. "Food and Social Relations at Nina Plantation." *American Anthropologist* 103: 671–691.

Shanks, Michael, and Christopher Tilley. 1982. "Ideology, Symbolic Power and Ritual Communication: A Reinterpretation of Neolithic Mortuary Practices." In *Symbolic and Structural Archaeology,* edited by I. Hodder, 129–154. Cambridge: Cambridge University Press.

Shennan, Stephen. 1978. "Archaeological 'Cultures': An Empirical Investigation." In *The Spatial Organization of Culture*, edited by I. Hodder, 113–139. London: Duckworth.

Sofaer, Joanna R. 2006a. *The Body as Material Culture: A Theoretical Osteoarchaeology.* Cambridge: Cambridge University Press.

———. 2006b. "Gender, Bioarchaeology and Human Ontogeny." In *Social Archaeology of Funerary Remains*, edited by R. Gowland and C. Knüsel, 155–167. Oxford: Oxbow Books.

Sofaer Derevenski, Joanna. 1994. "Where Are the Children?: Accessing Children in the Past." *Archaeological Review from Cambridge* 13: 7–20.

———. 1997a. "Age and Gender at the Site of Tiszapolgár-Basatanya, Hungary." *Antiquity* 71: 875–889.

———. 1997b. "Engendering Children, Engendering Archaeology." In *Invisible People and Processes: Writing Gender and Childhood into European Archaeology*, edited by J. Moore and E. Scott, 192–202. Leicester: Leicester University Press.

———, editor. 2000. *Children and Material Culture*. London: Routledge.

Solecki, Ralph S. 1971. *Shanidar: The First Flower People*. New York: Alfred A. Knopf.

Stirland, Ann. 1991. "Diagnosis of Occupationally Related Paleopathology: Can It Be Done?" In *Human Paleopathology: Current Syntheses and Future Options*, edited by D. J. Ortner and A. C. Aufderheide, 40–47. Washington, D.C.: Smithsonian Institution Press.

Tarlow, Sarah. 1999. *Bereavement and Commemoration: An Archaeology of Mortality.* Oxford: Blackwell.

Tiesler, Vera, and Andrea Cucina, editors. 2007. *New Perspectives on Human Sacrifice and Ritual Body Treatments in Ancient Maya Society*. Berlin: Springer-Verlag.

Tilley, Christopher. 1984. "Ideology and the Legitimation of Power in the Middle Neolithic of Southern Sweden." In *Ideology, Power, and Prehistory*, edited by D. Miller and C. Tilley, 111–146. Cambridge: Cambridge University Press.

Turner, Victor W. 1995 [1969]. *The Ritual Process: Structure and Anti-Structure*. New York: Aldine De Gruyter.

Vlahogiannis, Nicholas. 1998. "Disabling Bodies." In *Changing Bodies, Changing Meanings: Studies on the Human Body in Antiquity*, edited by D. Montserrat, 13–36. London: Routledge.

Voss, Barbara L., and Robert A. Schmidt. 2000. "Archaeologies of Sexuality: An Introduction." In *Archaeologies of Sexuality*, edited by R. A. Schmidt and B. L. Voss, 1–32. London: Routledge.

Walker, Phillip L., and Della Collins Cook. 1998. "Brief Communication: Gender and Sex: Vive la Difference." *American Journal of Physical Anthropology* 106: 255–259.

Wallace, Anthony F. C. 1966. *Religion: An Anthropological View*. New York: Random House.

Winters, Howard D. 1968. "Value Systems and Trade Cycles of the Late Archaic in the Midwest." In *New Perspectives in Archaeology*, edited by S. R. Binford and L. R. Binford, 175–222. Chicago: Aldine.

Wobst, H. Martin. 2000. "Agency in (Spite of) Material Culture." In *Agency in Archaeology*, edited by M.-A. Dobres and J. Robb, 40–50. London: Routledge.

PART I

Community Identity
and Ethnogenesis

3

Bridging Histories

The Bioarchaeology of Identity in Postcontact Florida

CHRISTOPHER M. STOJANOWSKI

Many modern Native American tribes, particularly in the eastern United States, are the products of complex social, political, and biological processes of adaptation that began with European contact in the last decade of the fifteenth century. Modern tribal identities such as Creek, Choctaw, and Seminole emerged during the intervening centuries and represent examples of historical ethnogenesis—the formation of new identities that were often unique, whose structural and cultural forms never existed previously, and that differed considerably from the ancestral "contributing" communities (Ferguson and Whitehead 1992; Moore 1994a, 1994b, 2001; Terrell 2001). It is now abundantly clear that the tribal formations of the postcontact Americas fit this description (see Hill 1996a and references therein; Albers 1993; Albers and James 1986; Moore 1987, 1994a, 1994b, 2001; Sharrock 1974). Beyond the historical interest, however, research on colonial ethnogenetic processes provides a framework for understanding the relationship between national and tribal boundaries and interethnic conflict in polyethnic social contexts. To understand the roots of ethnic violence we must first understand the circumstances under which group distinctiveness emerged. Here bioarchaeological perspectives, with their attendant temporal depth, are integral.

Bioarchaeology contributes an important diachronic perspective on the process and mechanisms of identity formation and supplements the considerable attention that ethnogenesis has received in the historical ethnographic (e.g., Hill 1996a) and ethnoarchaeological literatures (for contrasting opinions, see Collard et al. 2006 and Terrell et al. 1997). Biological anthropology, specifically human microevolutionary research, has not thus far contributed significantly to this body of theory.[1] This is unfortunate, because intermarriage is often presented as one component leading to ethnic merger (Albers 1996; Hickerson 1996). This lack of engagement by bioanthropologists may be understandable, however. Those working with modern DNA are limited to temporally restricted research designs on modern populations (see Nystrom, this volume), while those working with ancient DNA suffer from such limited sample sizes that large-scale population genetic interpretations are problematic. Morphological approaches using archaeological skeletal populations entail neither deficiency and are uniquely positioned to advance understanding of diachronic ethnogenetic pattern and process.[2]

In this chapter I present an analysis of population genetic variability and interpopulation distances among a series of indigenous communities in Spanish colonial Florida. I combine biodistance data with archaeological and historical information to argue that the Seminole, generally considered eighteenth-century migrants to Florida, are, in part, biological descendants of the Pre-Columbian population of the state whose ancestral connections to the landscape were erased by the vagaries of colonial ethnonymy (see Sattler 1996; Wickman 1999). Furthermore, I argue that these biological connections are ultimately responsible for the divergence of Seminole and Creek tribal identities, as currently recognized. This bioarchaeological analysis advances anthropological knowledge in two primary ways: (1) it documents, for the first time, the emergence of a new ethnic identity based on biological signatures without invoking an explicitly sociobiological framework (see van den Berghe 1981); and (2) it positions the Seminole as a direct descendant community in Florida with repercussions for cultural patrimony and heritage (see Wickman 1999). Both of these contributions speak to broader issues with contemporary relevance for legal proceedings as well as indigenous historical traditions.

I begin by discussing ethnogenetic theory as a unifying approach within anthropology. After developing a predictive model of group interaction patterns based on historical ethnographic data, I discuss the general history of Spanish La Florida from the sixteenth through the early eighteenth centuries and contrast this with received wisdom about the origins of the Georgia Creek and Florida Seminole. I then present an analysis of population genetic variability for La Florida populations for three periods and link these results to historical and archaeological data on the earliest Seminole migrants to Florida. What emerges is a biosocial narrative that bridges the two distinct periods that structure general historical perspectives on Florida's indigenous communities: pre-Seminole and Seminole.

Ethnogenetic Theory

Since William Sturtevant (1971) introduced the concept of ethnogenesis to American anthropology, interest in this topic has been robust, particularly in historical ethnography and archaeology. Distinctions in the two literatures are apparent, however, reflecting differences in the temporal scope of respective research designs and resulting differences in inferential coarseness. Ethnographers use historical texts, oral histories, and archaeological data to construct thick descriptive historical narratives that document the "emergence of a people . . . in relation to a sociocultural and linguistic heritage" (Hill 1996b: 1). Examples of this literature include Jonathan Hill (1996a), Barbara Kopytoff (1976), Patricia Albers (1993), and Gerald Sider (1994).

In contrast, the archaeological literature has two foci that are not completely congruent with that of historical ethnography: (1) determining the applicability of ethnogenetic theory in the remote past; in other words, the uniformitarian application of ethnogenetic models beyond the colonial period (Moore 1994a, 1994b, 2001; Terrell 2001), and (2) understanding the importance of ethnogenesis (as opposed to phylogenesis) as a mechanism of cultural transmissibility as related to evolutionary models of archaeological research (e.g., Collard et al. 2006). The distinctions here are subtle yet important. Ethnogenesis, formally defined, is simply the process by which new ethnic groups emerge, which is the primary interest of historical ethnographers. In exploring this process for postcolonial indigenous populations, however, it became apparent that phylogenetic models of cultural and biological co-evolution (cladistic branching) do not explain direct ethnographic observation (Moore 1994a, 1994b, 2001). Rather, historians documented complex cultural and biological relationships that defied clear categorization into monolithic cultural units (e.g., Albers 1993; Kopytoff 1976; Moore 1987; Quinn 1993; Sharrock 1974). "Tribes" were not internally homogenous entities but were polyglot, polyethnic, ephemeral formations with highly permeable social and geographic boundaries (Quinn 1993; Wolf 1982). Application of evolutionary models based on analogy with the static boundaries of biological species may therefore not be satisfactory (Moore 1994a, 1994b, 2001; Terrell 2001). From an archaeological perspective, ethnogenetic theory provides two novel contributions that affect models of prehistory and human action: (1) that human societies are, and always were, interconnected in complex ways (Lesser 1961; Wolf 1982), mirroring the predictions of the "Multiregional Model" of modern human emergence, and (2) as a result of this, that human societies periodically reorganize themselves into new social forms that are distinct from those of antecedent communities and not parsed along strict ethnic, cultural, or biological lines, with postcolonial Native American formations such as the Creek and Seminole representing perfect examples (see Moore 1994a, 1994b, 2001).

In this chapter I am concerned with the process of ethnic fusion, not modes of cultural trait transmission, and therefore draw theoretical models from the ethnographic literature. Both vertical and horizontal schemes have been presented (Albers 1996; Ferguson and Whitehead 1992; Hickerson 1996; Sider 1994; Whitehead 1992): the former describe the process of ethnogenesis in stages, while the latter define the types of new social forms that resulted from this process. Among the former, Nancy Hickerson's (1996: 70) model is most general. She developed a three-phase model likened to "life-cycle transitions." The initial phase involves separation or "the negation or severing of . . . existing group loyalties"; during the liminal phase (*sensu* van Gennep 1960), "surviving—usually dysfunctional—social and/or economic ties wither away, and alternative connections are initiated or strengthened"; and reintegration occurs when a "new identity is consolidated, affirmed through ritual and the adoption of a validating mythology" (Hickerson 1996: 70). Patricia Albers (1996) stresses the continuum of organizational forms that result from colonial ethnogenetic processes, noting the combination of nonlinear fissioning and mergers that exist within tribal zones; similar typologies have been presented by Gerald Sider (1994), R. Brian Ferguson and Neil Whitehead (1992), and Whitehead (1992).

A Brief History of Postcontact Florida

The First Spanish Period in Florida begins with the establishment of St. Augustine in 1565 by Pedro Menéndez de Avilés and officially ends with the 1763 annexation of St. Augustine by the British. As part of their strategy of pacification, the *conquista de almas* (conquest of souls), Spain enlisted Franciscan missionaries to pacify and convert indigenous populations. Between 1573 and 1650 a series of missions was established in the area around St. Augustine, north along the Florida and Georgia coast, and west through north-central Florida into the eastern panhandle near present-day Tallahassee (figure 3.1) (Gannon 1965; Geiger 1937; Lanning 1935; Oré 1936; Thomas 1990). This large region was not culturally homogenous. The Spanish recognized three major divisions that were used to divide the indigenous populations into administrative provinces: Apalachee, Guale, and Timucua

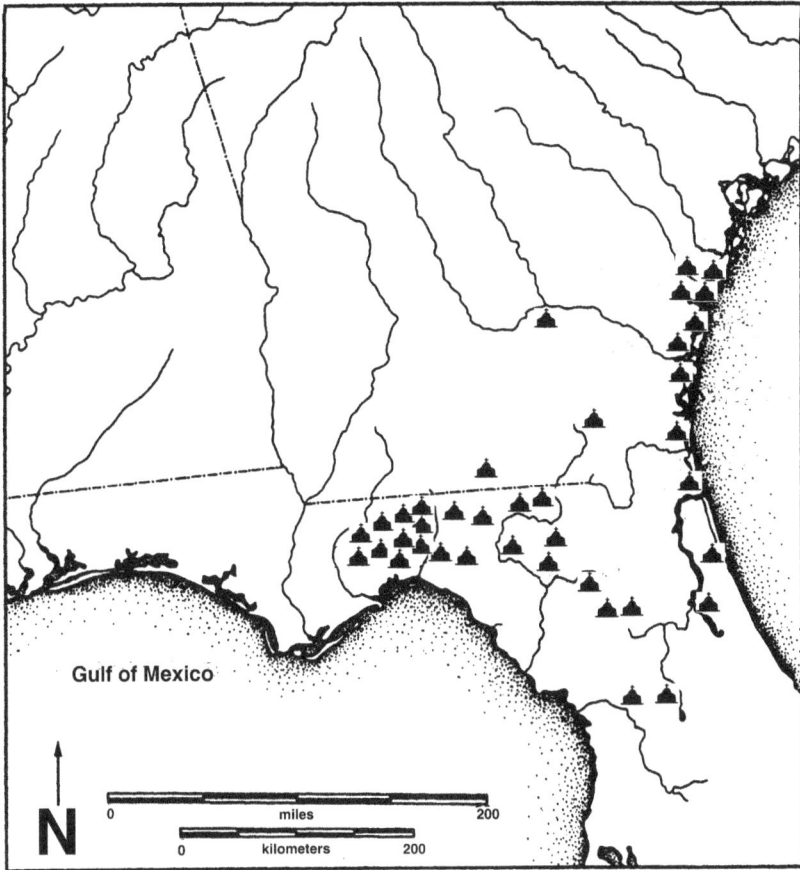

Figure 3.1. Map of La Florida with approximate mission locations noted (figure modified after Larsen 2001: fig. 2.1).

(figure 3.2). This tripartite typology continues to anchor modern scholarship's perspectives on Florida's past, and by most accounts these groups were linguistically, culturally, and politically distinct, with considerable internal differentiation. Approximately 25 languages or dialects were spoken among perhaps 50 different chiefdoms, which varied in population density, material culture, subsistence adaptation, and settlement structure (major syntheses of these cultural groups have been presented in Deagan 1978; Hann 1988, 1996a; Jones 1978; McEwan 2000; Milanich 1978, 1996, 2000, 2004; Saunders 2000; Worth 2004).

The mid-seventeenth century is generally considered the period of peak mission expansion (Geiger 1937), after which demographic collapse due to epidemic mortality and increased morbidity, reduced fertility, fugitiv-

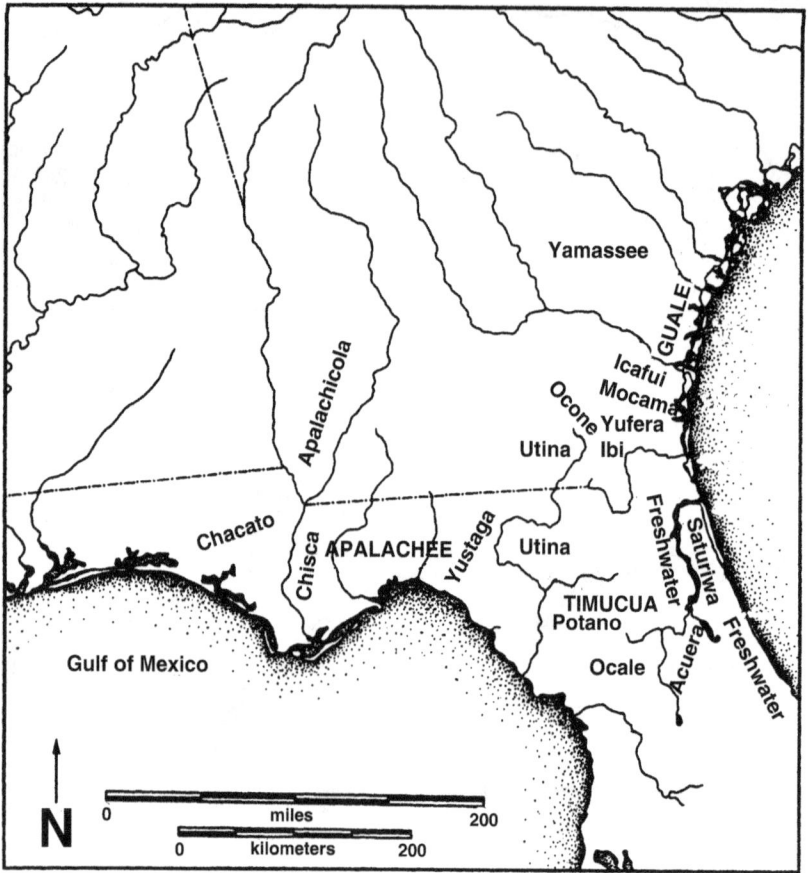

Figure 3.2. Map of tribal and ethnic groups in Florida and Georgia during the seventeenth and eighteenth centuries. Names in capital letters indicate the three administrative provinces adopted by the Spanish.

ism and out-migration, and slave raiding initiated a period of population aggregation and reorganization (Larsen 2001). Historian John Worth has documented this process of demographic restructuring for the Guale and Mocama living along the Atlantic coast (Worth 1995) as well as the western Timucua living in north-central Florida (Worth 1998a, 1998b). The process of aggregation was geographically hierarchical, proceeding from the local to the regional to the supraregional. The general scheme involved initial village aggregation to the local *doctrina* (mission church), the aggregation of formerly distinct *doctrinas*, and finally wholesale population replacement at missions that served a strategic function. Both local, converted Christian populations and migrant, non-Christian populations were used for this purpose (see Hann 1988: 102, 103, 165, 173–174; Worth 1995, 1998a, 1998b).

Although demographic issues were a primary cause of the collapse of the mission system, the proximate cause was burgeoning English interest in North America, whose plantation economy required a steady stream of labor (Hahn 2002; Smith 2002). Unlike the Spanish, who restricted firearms distribution, English traders armed their allies and actively encouraged slave raiding of Spanish villages and missions as part of what ethnographer Jonathan Hill (1996b: 5) would consider a "Hobbesian 'war of all against all.'" Slaving activity increased around 1660 (Bowne 2005; Covington 1968; Rountree 2002; Worth 1995) and intensified even further after the founding of Charlestown in 1670. Creek, Yamassee, Chisca, and Westo all participated in these assaults on the Spanish missions. Finally, after nearly a century of hardship that witnessed the reduction of Apalachee, Timucua, and Guale populations to fewer than 10,000 individuals (from an estimated 150,000 to 200,000 individuals—see the population size discussion in Hann 1988, 1996a; Milanich 1999; Stojanowski 2005a, 2005b; Worth 1995, 1998b), Colonel James Moore initiated a series of direct assaults on the Florida missions (Boyd et al. 1951; Thomas 1990). By 1706 the entire system had collapsed, the missions were razed or abandoned, and the remnant populations were taken as slaves, escaped west to French territory, or returned to St. Augustine, remaining Spanish loyalists until this city, too, was annexed by the English in 1763 (Covington 1964; Hann 1988).

To say that the Spanish missions were destroyed at the hands of the English provides only an incomplete version of events, however. Many members of the attacking army were, in fact, Creek who had allied themselves with the English and were actively involved in the deerskin and slave trade (Hann 1988; Knight 1994; Worth 2000). The history of the Creek is extremely complex and well beyond the scope of this chapter (see recent summaries in Knight 1994; Worth 2000). Labels such as the "Creek Confederacy" and their designation as one of the "Five Civilized Tribes," however, imply a degree of internal homogeneity that never existed in historic times. Rather, the term "Creek" referred to numerous semi-independent towns composed of remnant chiefdoms and displaced peoples ravaged by disease that congregated along the interior rivers of central Georgia and Alabama during the mid- to late seventeenth century. There is no doubt that some of the Creek were fugitives from the Spanish mission provinces (Hann 1988; Wickman 1999); the Georgia interior served as a refuge for discontents as early as the late sixteenth century (Oré 1936). As community defense was the *raison d'être*, the Creek should be considered a classic tribal confederacy (*sensu* Albers 1996; Sider 1994) whose name was bestowed by their English trading partners. They were Indians living along the numerous creeks in central Georgia, and the term "Creek Indians" was later simply shortened to "Creek."

During the first decades of the eighteenth century, some Lower Creek populations migrated down the Flint and Chattahoochee rivers into Spanish Florida (Sattler 1987, 1992, 1996; Sturtevant 1971; Weisman 1992, 2000). Both push and pull mechanisms explain this migration, including the Yamassee War of 1715, which witnessed Cherokee betrayal of the Creek and destroyed relationships with English traders at Charlestown (push), as well as desperate Spanish attempts to vivify their colony (pull) (reviewed in Weisman 2000). Further detachment from the Creek political body resulted from the geographic separation caused by this migration back to Spanish lands. Social and political isolation ensued, and the unique experience of living in Florida during the eighteenth century, along with the actions of specific historical figures and the provisions of specific treaties, resulted in the establishment of the two distinct tribal identities that we recognize today: Creek and Seminole (Covington 1993; Sturtevant 1971; Weisman 1992, 2000; Wright 1986).

Most agree, however, that it is the geographical separation of proto-Seminole and Creek settlements that was ultimately responsible for the development of distinct Creek and Seminole identities. This is the most critical point. The ultimate cause of Seminole ethnogenesis must be sought in their initial motivations for the emigration from Georgia. I propose that these motivations are manifested during the Spanish mission period and are directly related to the extent of out-migration documented during the seventeenth century and to changes in social organization that occurred as a consequence of demographic collapse and the global power struggle between England and Spain over control of eastern North America.

Analyzing Population Genetic Signatures

To analyze patterns of genetic diversity during the mission period, data were recorded for sixteen mesiodistal and buccolingual odontometric variables for approximately 1,200 individuals from 26 distinct archaeological samples (see Stojanowski 2001, 2005a: tables 1, 2, 3). These samples represent all three mission provinces for the following three periods: late precontact (AD 1300–1500, 19 samples), early mission (AD 1600–1650, Santa Catalina de Guale, ossuary at Santa Catalina de Guale de Santa María, San Martín de Timucua, San Pedro y San Pablo de Patale), and late mission (AD 1650–1704, Santa Catalina de Guale de Santa María, Santa María de los Yamassee, San Luis de Talimali). Precontact samples were combined into regional aggregate samples to increase sample sizes and counteract differential mortuary "catchment areas" (Stojanowski 2001, 2003, 2005b). These include panhandle Florida (Apalachee), central Florida (Timucua), coastal Georgia (Guale), and the two subsamples at Irene Mound, Georgia. Further details of site provenience can be found in previous publications (Stojanowski 2001, 2003, 2005a, 2005b).

The quantitative genetic statistical software package RMET was used to generate two population genetic parameters, F_{ST} and intersample genetic distances (Relethford 2003; Relethford et al. 1997). F_{ST} is a statistic that represents the overall level of genetic variation that exists within a regional mating network. This statistic varies between 0 and 1, where 0 indicates that all populations are genetically identical with no mating structure or genetic isolation and 1 indicates that all populations are genetically distinct with no genetic exchange between them. For this analysis, the magnitude of F_{ST} is less important than the change in F_{ST} through time; an increase in F_{ST} suggests that population sizes declined and/or gene flow was more restricted, and vice versa. The second parameter of interest is the pattern of genetic distances between populations. Whereas F_{ST} indicates the overall level of genetic diversity within a regional mating network, genetic distances indicate the pattern of relationships between each sample. RMET analyses were performed with a heritability of 0.62 (see Stojanowski 2005a, 2005b), and all distance matrices were bias corrected and scaled to reflect differential effective population sizes (see Stojanowski 2005a for these estimates; also see Relethford 2003 and Relethford et al. 1997 for discussion of bias correction and scaling).

Results

Phenotypic F_{ST} for the five aggregate precontact samples was 0.018 (*se* = 0.009), which differs significantly from 0 (*p* = 0.045) but is not excessive in magnitude considering the degree of linguistic and cultural diversity documented by historic chroniclers (see Barcia 1951; Bourne 1922; Vega 1951; also Deagan 1978; DePratter 1991; Hann 1988, 1996a; McEwan 2000; Milanich 1978, 1996, 1999, 2000, 2004). This statistic establishes the baseline level of diversity but is not meaningful by itself.

1600-1650   1650-1704

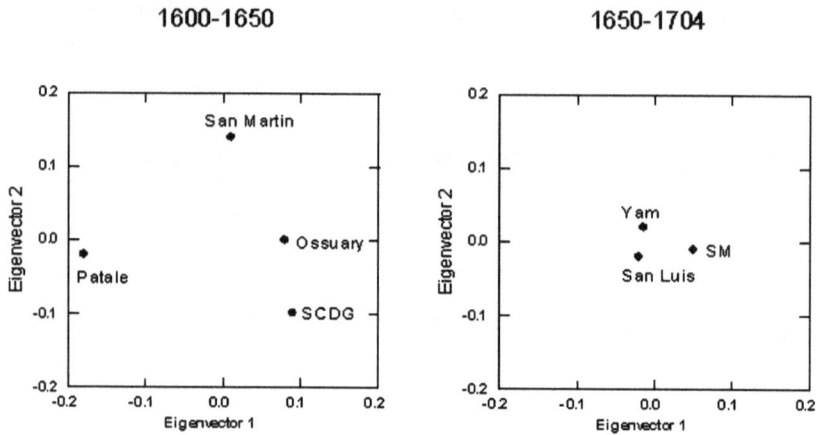

Figure 3.3. Genetic distance ordinations for the early mission period (AD 1600–1650) and late mission period (AD 1650–1704). Abbreviations are as follows: Patale = San Pedro y San Pablo de Patale, San Martín = San Martín de Timucua, Ossuary = Ossuary at Santa Catalina de Guale, SCDG = Santa Catalina de Guale, Yam = Santa María de los Yamassee, SM = Santa Catalina de Guale de Santa María, San Luis = San Luis de Talimali.

Phenotypic F_{ST} for the four early mission period samples (San Pedro y San Pablo de Patale, Santa Catalina de Guale, ossuary at Santa Catalina de Guale, and San Martín de Timucua) is larger in magnitude (F_{ST} = 0.024) and significantly different from 0 (p = 0.002). The pattern of genetic distances among samples during the early mission period indicates isolation by distance (figure 3.3), which reflects limited long-distance gene flow and suggests that the pattern of mate exchange did not change with the transition to settled mission life. This pattern of population genetic changes is most consistent with declining population size and decreased biological interaction among missionized populations during the immediate postcontact period.

Phenotypic F_{ST} for the three late mission period samples (San Luis de Talimali, Santa María de los Yamassee, Santa Catalina de Guale de Santa María) indicates a reversal in genetic variance trends. F_{ST} decreases to 0.002 (se = 0.0009), which is significantly different from 0 (p = 0.026), significantly smaller than the early mission period F_{ST} (p = 0.006), and marginally insignificant in relationship to the late precontact period F_{ST} (p = 0.075). The genetic distance ordination indicates a lack of biological structure, with no evidence for isolation by distance (figure 3.3). This result is unexpected given the known changes in population size that occurred throughout the seventeenth century. If population size alone were driving the genetic variances, late mission period regional genetic diversity would be maximized. This is clearly not the case. Rather, this analysis indicates that something very different was occurring after 1650. This, I propose, is extensive long-distance migration and gene flow among missionized communities in the wake of near-complete demographic collapse.

Discussion

To interpret these data they must be incorporated into an appropriate historical framework to explore how social forces defined the two major transitions considered in this analysis: the transition to life as Spanish Catholics living at the missions (early mission period) and the transition to a post-demographic collapse tribal indigenous community (late mission period).

The early mission period witnessed significant changes in native society. Inclusion within the Spanish political sphere required cooperation and peace among Christian populations, despite the internecine conflict that characterized precontact southeastern community relationships (Bourne 1922; Laudonnière 1975; Vega 1951). I propose that the lack of small-scale intergroup warfare was partially responsible for the increase in regional genetic heterogeneity, reflecting the fact that warfare was a biologically integrative process in which the capture of women and children was a consequence, if not cause, of intergroup violence (DePratter 1991; Laudonnière 1975: 27; Smith 1987). In addition to warfare, other forms of intertribal integration must have been disrupted, such as partial cohabitation for trading, barter, communal subsistence and ceremonial activities, and mergers for defense alliances (Quinn 1993). In accordance with generalized models of state expansion (Ferguson and Whitehead 1992), conflict did not disappear from the lives of the indigenous communities. Rather, the structure of conflict was redefined during the early mission period and took the form of indigenous revolts directed against the Spanish. Violent uprisings before 1650 were often ascribed to local populations, with no evidence for a widespread, pan-ethnic uprising that united all social groups against the Spanish Crown.[3] This pattern of conflict, I propose, reflects tension internal to the Spanish system caused by demographic collapse, changes in social organization, and the resulting tribalization of communities competing for perceived or actual resources (Ferguson and Whitehead 1992). The structure of political interactions was local in scale, because English interests in the mid-Atlantic were nascent, and contained within the Spanish portion of the "tribal zone" in La Florida. This phase of colonial transformation represents not a merger but short-term fissioning of community organization as a result of ethnic factionalism (Hill 1996b) and is consistent with Nancy Hickerson's (1996) "separation" phase of ethnogenesis. Precontact systems of intertribal integration were disrupted and patterns of economic co-dependency were shattered. Local populations struggled to establish a footing in this new sociopolitical system that created a tribal zone of exploratory social adaptation at the fringe of the Spanish empire.

The dramatic reduction in population genetic variability and disappearance of isolation by distance among late precontact period samples indicate that a single biological population was resident in La Florida, reflecting social adaptation in a post–demographic collapse context. By 1650 the coastal Guale and eastern Timucua had been reduced to less than 5% of their precontact numbers, while the western Timucua and Apalachee were in the active process of demographic collapse (Hann 1988, 1996a; Milanich 1999, 2000; Worth 1995, 1998b). Spanish and native responses to this demographic crisis were synergistic and represent factors both extrinsic and intrinsic to native society. The Spanish initiated an aggressive campaign of population reorganization in which nuances of ethnic and linguistic diversity were overlooked and non-Christian populations were welcomed under the Spanish banner (see examples in Hann 1988: 102, 103, 165, 173, 179). As a testament to Spanish desperation and the confused nature of group allegiances, some of these pagan populations were actually raiding other Spanish missions in different provinces (see references to the Chisca and Yamassee in Covington 1968; Hann 1988; and Worth 1998b). Indeed, the Spanish mode of conversion, which entailed the distribution of gift and prestige items, drew many nonlocal populations to the fringe of the Spanish empire (Rountree 2002; Smith 2002; Waselkov 1989). Migration between provinces was also documented (e.g., Hann 1986b: 102, 1996a: 25; Worth 1995, 1998b) and was particularly intense after 1650 and the interior upheaval in the aftermath of the Timucua rebellion in 1656 (Worth 1998a, 1998b). People were on the move.

Long-distance migration initiated a motive of opportunity for geographically expansive mate exchange as formerly distinct populations were redistributed across the landscape. Another consequence of demographic collapse, I propose, was that local villages, in some provinces, were no longer functional and traditional systems of mate exchange at the local community level were no longer tenable. This, when combined with rampant fugitivism often in the form of male transiency (e.g., Bushnell 1979: 5; Hann 1986a: 379, 1988: 171), created motives of reproductive necessity demonstrably requiring a broadening of the networks of mate exchange. In Patricia Wickman's analysis, such flexibility in group composition, mating patterns, and residences was entirely consistent with Maskókî cosmogony. Again, it is difficult to define formally what this level of community size may have been; however, this interpretation is supported by documented interethnic marriages (e.g., Hann 1988: 184) and formal complaints of village size functionality (e.g., Bushnell 1979: 5; Hann 1986b: 103, 1988: 171; Worth 1998b: 32), which predominate during the post-1650 period.

Thus far I have demonstrated that early mission period populations were actively decreasing in size but had not done so to the point where long-distance migration and changes in traditional systems of social organization were needed. This period witnessed initial conversion to a Catholic lifestyle, with resulting increases in between-group variability due to changes in social structures that previously led to intertribal biological integration. The late mission period represented a post–demographic collapse environment in which reduced local population sizes led to increased long-distance migration, social boundary permeability, and social adaptations that restructured Native American mating behavior. This pattern of biosocial change is consistent with Hickerson's (1996: 70) second phase of ethnogenesis, liminality, in which "dysfunctional social ties wither away and alternative connections are initiated." This reflects the development of a "hybridized group coalition" (*sensu* Albers 1996: 93) in which social stress in a post–demographic collapse context required individuals to "widen the basis of their ethnicity to find a lowest common denominator" (Kopytoff 1976: 34) as part of a community-wide effort to survive. Similar processes were at the heart of Maroon ethnogenesis (Bilby 1996; Kopytoff 1976), in which peoples "minimize the cultural saliency of their differences in order to create some level of common identity and intercourse" (Albers and James 1986: 12). At this point in the ethnogenetic process a tributary form of organization (*sensu* Sider 1994) is evident.

Although population size was certainly an important catalyst for microevolutionary change, demography alone does not explain the *active* participation of indigenous communities in this process. Indeed, the impetus for this emergent ethnic identity was on a more global scale, reflecting the competitive European zeitgeist and the omnipresent tool of warfare and conflict (Ferguson and Whitehead 1992). At about the same time when demographic collapse was nearly complete for most of the eastern Spanish provinces (ca. 1650), indigenous revolts directed against the Spanish all but ceased. Simultaneously, the presence of the English in the southeastern

United States initiated widespread aboriginal adaptations to the competing European colonization models and economies (Rountree 2002; Sider
1994). The Creek and Westo, for example, became predatory slave raid-
ers, targeting the unarmed Spanish mission populations located in Florida
and Georgia (Bowne 2005; Knight 1994). As slaving intensified, raids from
within La Florida were launched into the hinterland; the results were often
disastrous (e.g., Covington 1967), reflecting differential English and Spanish policies toward the provisioning of firearms. Conflict once again transformed into an indigenous-indigenous structure, pitting "Spanish-Catholic
Indians" against armed, English-allied, predatory tribal formations of the
Georgia and Carolina interior. In this context of ethnocide, local tensions
assumed global proportions. Ethnic differences were minimized and, as
demonstrated here, mate exchange became more widespread, as a new ethnic collective emerged in La Florida. What is most interesting is that this
liminal identity was signaled biologically in the absence of any corresponding social homogenization (see Deagan 1983; Worth 1995: 4); old ethnic or
tribal labels persisted unfettered. This population was unrecognized and
therefore unnamed by contemporary observers and was ultimately subject
to widespread diaspora when the English destroyed the Spanish mission
system in the first decade of the eighteenth century. However, I argue that
this network of kin relations represents a nascent biological population,
with a nascent identity of "Spanish-Catholic Indian" that emerged in the
wake of demographic collapse and in the context of competing Spanish and
English colonies that formed a southeastern U.S. tribal zone.

Bridging Histories: Seminole Ethnogenesis Reconsidered

To understand how this relates to Seminole ethnogenesis two additional
factors must be considered. The first factor is the extent of out-migration
and fugitivism that was occurring throughout the seventeenth century and
its effect, in concert with disease, on declining population size in mission
communities. My previous reading of the biological data (namely, paleopathology and genetic variance estimates: Stojanowski 2001, 2005b) suggests
that the effects of epidemics on population demography, while certainly of
primary importance, may have overshadowed the more general "slow leak-
ing" of populations to the Georgia interior. How many Apalachee, Guale,
and Timucua were already living in the Creek heartland by the end of the
seventeenth century, and (given the analyses presented here) how interbred
and polyethnic was the Georgia interior even before anything resembling a

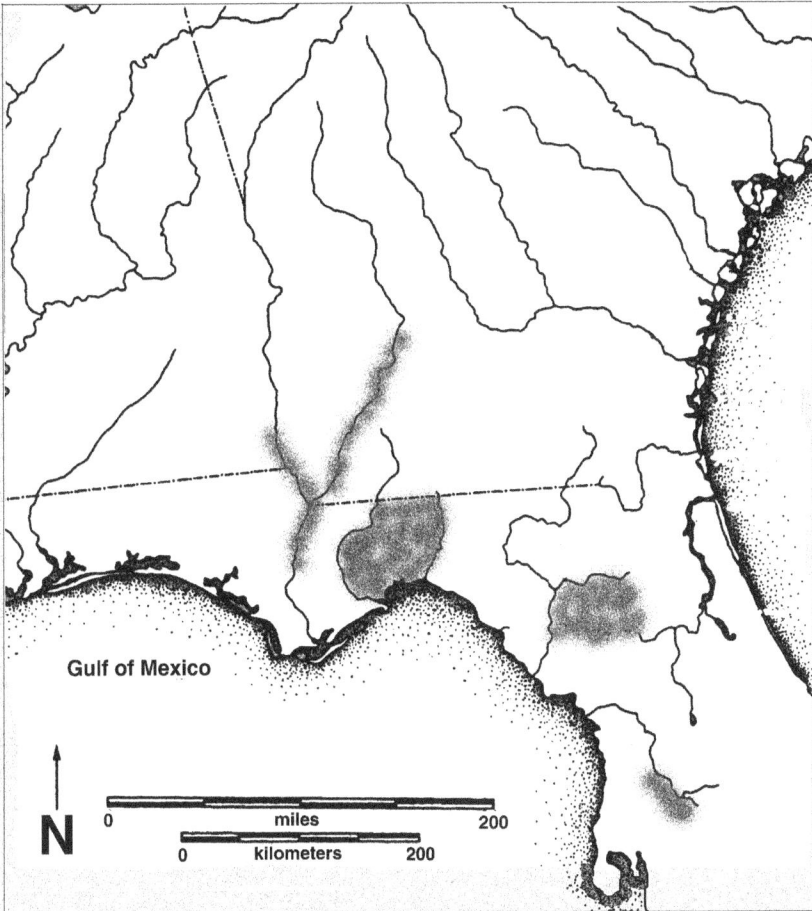

Figure 3.4. Highlighted areas represent Seminole settlement areas between AD 1720 and AD 1784 (figure modified after Sattler 1987: 22).

Creek political body existed? I am not alone in inferring these relationships, as Spanish period linguistic (Hann 1996b) and ceramic data suggest similar connections (Worth 2000).

The second factor is the physical location of the earliest proto-Seminole settlements in Florida, as documented historically (figure 3.4) (Sattler 1987). Comparison of figure 3.4 with the locations of late seventeenth century mission population density reveals remarkable similarities. The Tallahassee Red Hills (Apalachee) and Alachua Prairie (western Timucua) were, for reasons still unknown, the last and least to suffer the effects of demographic collapse

(Hann 1986a, 1988; Worth 1998b). Interestingly, the two areas were also re-settled by completely distinct proto-Seminole populations that remained separate for much of their history (Weisman 1992, 2000). Therefore, when I speak of the Seminole I am referring to those populations that resided in the Tallahassee Red Hills, where the Apalachee, who represented over 90% of the surviving Spanish mission population by 1706, resided (Hann 1988: 164).

Although these locations may have presented attractive grazing and/or farming opportunities (Craig and Peebles 1974), I propose a far more agent-based explanation for directed resettlement of Apalachee province. Simply put, the choice of settlement location represents the political influence of the elite Christian, pro-Spanish elements that had intermarried with "Creek" populations beginning around 1650. And in this regard the Apalachee are key. They were feared and respected throughout the region (see Bourne 1922), they maintained comparatively large population sizes throughout the seventeenth century (Hann 1988: 164), they were widely dispersed after the mission period (Hann 1988; Boyd et al. 1951), their territory represents one of the locations where proto-Seminole settlements were first established in the mid-eighteenth century (Sattler 1987, 1992, 1996), and the elites involved in repopulating Apalachee were married to Apalachee women (Emperor Brimms and his son Secoffee were both married to Apalachee) (see Fair-banks 1978: 164; Hann 1996b: 67; Sattler 1987, 1992, 1996).

Therefore, the move back to Apalachee territory by proto-Seminole/Lower Creek populations is in my opinion simply a return to the homeland of displaced peoples, renamed by contemporary Europeans that failed to understand the nuances of colonial tribal society. If Seminole emergence ultimately resulted from the physical and social separation of this move back to Spanish Florida, then the foundations of Seminole ethnogenesis are not to be found in the early eighteenth century archaeological or historical records but in the mid-seventeenth century, when expanding mate exchange networks were redefining the social fabric of native communities throughout the region. Following Hickerson (1996), the separation and liminal stages of Seminole ethnogenesis pass in rapid succession, being completed circa 1650. That initial stages of social adaptation occurred rapidly and in quick succession is intuitive; the alternative is ethnocide. Reintegration (that is, the consolidation of Seminole identity and the reification of a common mythology) would take far longer and is arguably still ongoing today (see Weisman 1992 and Wickman 1999 on core element conservatism).

Conclusion

Combining historical, archaeological, and biological perspectives on Florida's indigenous communities results in a more complete and nuanced biosocial narrative of biocultural adaptation. By 1650 extensive alteration in Apalachee, Guale, and Timucua identity had become evident. Though still maintaining some semblance of distinct identities from the past, they were in a liminal stage and were becoming "Seminole." This transformation was prefaced by expanding networks of intermarriage, which redefined the nature of social identity in emerging polyethnic communities scattered throughout Florida and Georgia. Perhaps out of biological necessity (local village size), or perhaps due to new opportunities (fugitivism and migration), extensive gene flow across ethnic boundaries characterized the southeastern United States. Ties of kinship united disparate, sometimes geographically distant elites and commoners, which provided the thread of commonality, of shared identity, in times of turmoil that required expedient action. In the context of a three- and then four-way power struggle among European nations, intergroup conflict and generalized unrest were common, and migration provided a tried-and-true formula for ameliorating unfavorable living conditions. Ultimately it is this system of consanguinity that predicted, allowed, and foreshadowed the Creek Confederacy and the Seminole ethnogenesis that resulted from this. For the Seminole, intermarriage did not solidify a new identity with bonds of kinship; rather, interethnic gene flow was primary, the first step in forging this new identity. Shared experience, continued American encroachment, and ever increasing physical, social, and political isolation from the Creek Confederacy promoted this emerging identity. But it is the dissolution of precontact ethnic labels, resulting social restructuring, and widespread genetic integration documented in this chapter that initially helped promote the move to Florida.

Recognizing biological and social connections in the context of a coherent historical trajectory evident from the pre-Creek through Creek and then Seminole periods has important implications for the parsing of cultural patrimony in the southeastern United States. Labeling the Seminole as eighteenth-century migrants to Florida diminishes their patrimony, history, and heritage. Just as the Creek are a case study in colonial ethnonymy, so too are the Seminole. "Seminole" is a corrupted form of *cimarron* (Spanish for wild, untamed, and uncontrollable), a condition that defined Seminole resistance throughout their recognized history. I propose that this same passion for resistance further links the modern Seminole to Florida's precontact indigenous populations; those that "quit" the Spanish became Creek and ultimately navigated the colonial period, later defying the English and then the Americans. This history of defiance, which figures so prominently in Seminole identity today, includes surviving three wars with the United States and population reductions to 200 people. Defiance served the Seminole well throughout their history—and a long history it has been.

Notes

1. There is actually an extensive literature in anthropological genetics that correlates various types of social or political phenomena (such as linguistic, ethnic, religious, political barriers) with patterns of genetic variation. Although useful for evaluating the saliency of social barriers to mate exchange, these approaches are generally limited to synchronic data sets, and therefore they cannot directly examine the historical processes that resulted in the formation of the social phenomena (such as an ethnic group or tribal community) being investigated (e.g., Barbujani and Sokal 1990).

2. To be fair on all accounts, biodistance analyses using morphological data are hampered by an incomplete understanding of the genetic and environmental contributions to the phenotype.

3. The major revolts were in 1597 (Guale), 1647 (Apalachee), and 1656 (Timucua). Other revolts were recorded in 1576, 1587, 1608, 1645, and 1680 for the Guale (Hann 1988; Saunders 2000; Worth 1992, 1998b).

References Cited

Albers, Patricia C. 1993. "Symbiosis, Merger, and War: Contrasting Forms of Intertribal Relationship among Historic Plains Indians." In *The Political Economy of North American Indians*, edited by J. H. Moore, 94–132. Norman: University of Oklahoma Press.

———. 1996. "Changing Patterns of Ethnicity in the Northeastern Plains, 1780–1870." In *History, Power, and Identity: Ethnogenesis in the Americas, 1492–1992*, edited by J. D. Hill, 90–118. Iowa City: University of Iowa Press.

Albers, Patricia C., and William R. James. 1986. "On the Dialectics of Ethnicity: To Be or Not to Be Santee Sioux." *Journal of Ethnic Studies* 14: 1–27.

Barbujani, Guido, and Robert R. Sokal. 1990. "Zones of Sharp Genetic Change in Europe Are Also Linguistic Boundaries." *Proceedings of the National Academy of Sciences USA* 87: 1816–1819.

Barcia, Andres de. 1951. *Ensayo cronologico, para la historia general de la Florida*. Translated by A. Kerrigan. Gainesville: University of Florida Press.

Bilby, Kenneth. 1996. "Ethnogenesis in the Guianas and Jamaica: Two Maroon Cases." In *History, Power, and Identity: Ethnogenesis in the Americas, 1492–1992*, edited by J. D. Hill, 119–141. Iowa City: University of Iowa Press.

Bourne, Edward G., editor. 1922. *Narratives of the Career of Hernando de Soto in the Conquest of Florida*. 2 vols. New York: Allerton.

Bowne, Eric E. 2005. *The Westo Indians: Slave Traders of the Early Colonial South*. Tuscaloosa: University of Alabama Press.

Boyd, Mark F., Hale G. Smith, and John W. Griffin. 1951. *Here They Once Stood: The Tragic End of the Apalachee Missions*. Gainesville: University of Florida Press.

Bushnell, Amy T. 1979. "Patricio de Hinachuba: Defender of the Word of God, the Crown of the King and the Little Children of Ivitachuco." *American Indian Culture and Research Journal* 3: 1–21.

Collard, Mark, Stephen J. Shennan, and Jamshid J. Tehrani. 2006. "Branching, Blending, and the Evolution of Cultural Similarities and Differences among Human Populations." *Evolution and Human Behavior* 27: 169–184.

Covington, James W. 1964. "The Apalachee Indians Move West." *Florida Anthropologist* 17: 221–225.

———. 1967. "Some Observations concerning the Florida-Carolina Indian Slave Trade." *Florida Anthropologist* 20: 10–18.

———. 1968. "Stuart's Town, the Yamassee Indians and Spanish Florida." *Florida Anthropologist* 21: 8–13.

———. 1993. *The Seminoles of Florida*. Gainesville: University Press of Florida.

Craig, Alan K., and Christopher S. Peebles. 1974. "Ethnoecologic Change among the Seminoles, 1740–1840." *Geoscience and Man* 5: 83–96.

Deagan, Kathleen A. 1978. "Cultures in Transition: Fusion and Assimilation among the Eastern Timucua." In *Tacachale: Essays on the Indians of Florida and Southeastern Georgia during the Historic Period*, edited by J. T. Milanich and S. Proctor, 89–119. Gainesville: University Press of Florida.

———. 1983. *Spanish St. Augustine: The Archaeology of a Colonial Creole Community*. New York: Academic Press.

DePratter, Chester B. 1991. *Late Prehistoric and Early Historic Chiefdoms in the Southeastern United States*. New York: Garland Publishing, Inc.

Fairbanks, Charles H. 1978. "The Ethno-Archeology of the Florida Seminole." In *Tacachale: Essays on the Indians of Florida and Southeast Georgia during the Historic Period*, edited by J. T. Milanich and S. Proctor, 163–193. Gainesville: University Press of Florida.

Ferguson, R. Brian, and Neil L. Whitehead, editors. 1992. *War in the Tribal Zone: Expanding States and Indigenous Warfare*. Santa Fe: School of American Research Press.

Gannon, Michael V. 1965. *The Cross in the Sand: The Early Catholic Church in Florida, 1513–1870*. Gainesville: University of Florida Press.

Geiger, Maynard. 1937. *The Franciscan Conquest of Florida (1573–1618)*. Washington, D.C.: Catholic University of America.

Hahn, Steven C. 2002. "The Mother of Necessity: Carolina, the Creek Indians, and the Making of a New Order in the American Southeast, 1670–1763." In *The Transformation of the Southeastern Indians, 1540–1760*, edited by R. Etheridge, 79–114. Jackson: University Press of Mississippi.

Hann, John H. 1986a. "Demographic Patterns and Changes in Mid-Seventeenth Century Timucua and Apalachee." *Florida Historical Quarterly* 64: 371–392.

———. 1986b. *Spanish Translations*. Florida Archaeology No. 2. Tallahassee: Florida Bureau of Archaeological Research.

———. 1988. *Apalachee: The Land between the Rivers*. Gainesville: University Presses of Florida.

———. 1996a. *A History of the Timucua Indians and Missions*. Gainesville: University Press of Florida.

———. 1996b. "Late Seventeenth-Century Forebears of the Lower Creeks and Seminoles." *Southeastern Archaeology* 15: 66–80.

Hickerson, Nancy P. 1996. "Ethnogenesis in the South Plains: Jumano or Kiowa?" In *History, Power, and Identity: Ethnogenesis in the Americas, 1492–1992*, edited by J. D. Hill, 70–89. Iowa City: University of Iowa Press.

Hill, Jonathan D., editor. 1996a. *History, Power, and Identity: Ethnogenesis in the Americas, 1492–1992*. Iowa City: University of Iowa Press.

———. 1996b. "Introduction: Ethnogenesis in the Americas, 1492–1992." In *History, Power, and Identity: Ethnogenesis in the Americas, 1492–1992*, edited by J. D. Hill, 1–19. Iowa City: University of Iowa Press.

Jones, Grant D. 1978. "The Ethnohistory of the Guale Coast through 1684." In *The Anthropology of St. Catherines Island, 1: Natural and Cultural History*, edited by D. H. Thomas, G. D. Jones, R. S. Durham, and C. S. Larsen, 178–210. Anthropological Papers of the American Museum of Natural History, vol. 55, part 2. New York: American Museum of Natural History.

Knight, Vernon J. 1994. "The Formation of the Creeks." In *The Forgotten Centuries: Indians and Europeans in the American South, 1521–1704*, edited by C. Hudson and C. C. Tesser, 373–392. Athens: University of Georgia Press.

Kopytoff, Barbara. 1976. "The Development of Jamaican Maroon Ethnicity." *Caribbean Quarterly* 22: 33–50.

Lanning, John T. 1935. *The Spanish Missions of Georgia*. Chapel Hill: University of North Carolina Press.

Larsen, Clark S., editor. 2001. *Bioarchaeology of Spanish Florida: The Impact of Colonialism*. Gainesville: University Press of Florida.

Laudonnière, René. 1975. *Three Voyages*. Translated by C. E. Bannett. Gainesville: University Presses of Florida.

Lesser, Alexander. 1961. "Social Fields and the Evolution of Society." *Southwestern Journal of Anthropology* 17: 40–48.

McEwan, Bonnie G. 2000. "The Apalachee Indians of Northwest Florida." In *Indians of the Greater Southeast: Historical Archaeology and Ethnohistory*, edited by B. G. McEwan, 57–84. Gainesville: University Press of Florida.

Milanich, Jerald T. 1978. "The Western Timucua: Patterns of Acculturation and Change." In *Tacachale: Essays on the Indians of Florida and Southeastern Georgia during the Historic Period*, edited by J. T. Milanich and S. Proctor, 59–88. Gainesville: University Press of Florida.

———. 1996. *The Timucua*. Cambridge, Mass.: Blackwell Publishers.

———. 1999. *Laboring in the Fields of the Lord: Spanish Missions and Southeastern Indians*. Gainesville: University Press of Florida.

———. 2000. "The Timucua Indians of Northern Florida and Southern Georgia." In *Indians of the Greater Southeast: Historical Archaeology and Ethnohistory*, edited by B. G. McEwan, 1–25. Gainesville: University Press of Florida.

———. 2004. "Timucua." In *Handbook of North American Indians, Vol. 14: Southeast*, edited by R. Fogelson and W. Sturtevant, 218–228. Washington, D.C.: Smithsonian Institution Press.

Moore, John H. 1987. *The Cheyenne Nation: A Social and Demographic History*. Lincoln: University of Nebraska Press.

———. 1994a. "Ethnogenetic Theory." *National Geographic Research and Exploration* 10: 10–23.

———. 1994b. "Putting Anthropology Back Together Again: The Ethnogenetic Critique of Cladistic Theory." *American Anthropologist* 96: 925–948.

———. 2001. "Ethnogenetic Patterns in Native North America." In *Archaeology, Language, and History: Essays on Culture and Ethnicity*, edited by J. E. Terrell, 31–56. Westport, Conn.: Bergin and Garvey.

Oré, Luis G. 1936. *The Martyrs of Florida*. New York: Joseph F. Wagner.

Quinn, William W., Jr. 1993. "Intertribal Integration: The Ethnological Argument in *Duro v. Reina*." *Ethnohistory* 40: 34–69.

Relethford, John H. 2003. "Anthropometric Data and Population History." In *Human Biologists in the Archives: Demography, Health, Nutrition, and Genetics in Historical Populations*, edited by D. A. Herring and A. C. Swedlund, 32–52. Cambridge: Cambridge University Press.

Relethford, John H., Michael H. Crawford, and John Blangero. 1997. "Genetic Drift and Gene-Flow in Post-Famine Ireland." *Human Biology* 69: 443–465.

Rountree, Helen C. 2002. "Trouble Coming Southward: Emanations through and from Virginia, 1607–1675." In *The Transformation of the Southeastern Indians, 1540–1760*, edited by R. Etheridge, 65–78. Jackson: University Press of Mississippi.

Sattler, Richard A. 1987. "Seminoli Italwa: Socio-political Change among the Oklahoma Seminoles between Removal and Allotment, 1836–1905." Doctoral dissertation, University of Oklahoma, Department of Anthropology.

———. 1992. "Ethnic Transformation on the Gulf Coast: The Apalachi Case." Paper presented at the Annual Meeting of the American Association for the Advancement of Science, Chicago.

———. 1996. "Remnants, Renegades, and Runaways: Seminole Ethnogenesis Reconsidered." In *History, Power, and Identity: Ethnogenesis in the Americas, 1492–1992*, edited by J. D. Hill, 36–69. Iowa City: University of Iowa Press.

Saunders, Rebecca. 2000. *Stability and Change in Guale Indian Pottery, A.D. 1300–1702*. Tuscaloosa: University of Alabama Press.

Sharrock, Susan R. 1974. "Crees, Cree-Assiniboines and Assiniboines: Interethnic Social Organization on the Far Northern Plains." *Ethnohistory* 21: 95–122.

Sider, Gerald. 1994. "Identity as History: Ethnohistory, Ethnogenesis and Ethnocide in the Southeastern United States." *Identities* 1: 109–122.

Smith, Marvin T. 1987. *Archaeology of Aboriginal Culture Change in the Interior Southeast: Depopulation during the Early Historic Period*. Gainesville: University Press of Florida.

———. 2002. "Aboriginal Population Movements in the Postcontact Southeast." In *The Transformation of the Southeastern Indians, 1540–1760*, edited by R. Etheridge, 3–20. Jackson: University Press of Mississippi.

Stojanowski, Christopher M. 2001. "Cemetery Structure, Population Aggregation, and Biological Variability in the Mission Centers of La Florida." Doctoral dissertation, University of New Mexico, Department of Anthropology.

———. 2003. "Matrix Decomposition Model for Investigating Prehistoric Intracemetery Biological Variation." *American Journal of Physical Anthropology* 122: 216–231.

———. 2005a. "The Bioarchaeology of Identity in Spanish Colonial Florida: Social and Evolutionary Transformation before, during, and after Demographic Collapse." *American Anthropologist* 107: 417–431.

———. 2005b. *Biocultural Histories in La Florida: A Bioarchaeological Perspective.* Tuscaloosa: University of Alabama Press.

Sturtevant, William C. 1971. "Creek into Seminole: North American Indians." In *Historical Perspective*, edited by E. Leacock and N. Lurie, 92–128. New York: Random House.

Terrell, John E. 2001. "The Uncommon Sense of Race, Language, and Culture." In *Archaeology, Language, and History: Essays on Culture and Ethnicity*, edited by J. E. Terrell, 11–30. Westport, Conn.: Bergin and Garvey.

Terrell, John E., Terry L. Hunt, and Chris Gosden. 1997. "The Dimensions of Social Life in the Pacific: Human Diversity and the Myth of the Primitive Isolate." *Current Anthropology* 38: 155–195.

Thomas, David H. 1990. "The Spanish Missions of La Florida: An Overview." In *Columbian Consequences, Volume 2: Archaeological and Historical Perspectives on the Spanish Borderlands East*, edited by D. H. Thomas, 357–398. Washington, D.C.: Smithsonian Institution Press.

van den Berghe, Pierre L. 1981. *The Ethnic Phenomenon.* New York: Elsevier.

van Gennep, Arnold. 1960. *The Rites of Passage.* Chicago: University of Chicago Press.

Vega, Garcilaso de la. 1951. *The Florida of the Inca.* Translated by J. G. Varner and J. J. Varner. Austin: University of Texas Press.

Waselkov, Gregory A. 1989. "Seventeenth-Century Trade in the Colonial Southeast." *Southeastern Archaeology* 8: 117–133.

Weisman, Brent R. 1992. *Like Beads on a String: A Culture History of the Seminole Indians in North Peninsular Florida.* Tuscaloosa: University of Alabama Press.

———. 2000. "Archaeological Perspectives on Florida Seminole Ethnogenesis." In *Indians of the Greater Southeast*, edited by B. G. McEwan, 299–317. Gainesville: University Press of Florida.

Whitehead, Neil L. 1992. "Tribes Make States and States Make Tribes: Warfare and the Creation of Colonial Tribes and States in Northeastern South America." In *War in the Tribal Zone: Expanding States and Indigenous Warfare*, edited by R. B. Ferguson and N. L. Whitehead, 127–150. Santa Fe: School of American Research Press.

Wickman, Patricia R. 1999. *The Tree That Bends: Discourse, Power, and the Survival of the Maskókî People.* Tuscaloosa: University of Alabama Press.

Wolf, Eric. 1982. *Europe and the People without History*. Berkeley: University of California Press.

Worth, John E. 1992. "The Timucuan Missions of Spanish Florida and the Rebellion of 1656." Doctoral dissertation, University of Florida, Department of Anthropology.

———. 1995. *The Struggle for the Georgia Coast: An Eighteenth-Century Spanish Retrospective on Guale and Mocama*. American Museum of Natural History, Anthropological Papers, No. 75. New York: American Museum of Natural History.

———. 1998a. *The Timucuan Chiefdoms of Spanish Florida, Volume 1: Assimilation*. Gainesville: University Press of Florida.

———. 1998b. *The Timucuan Chiefdoms of Spanish Florida, Volume 2: Resistance and Destruction*. Gainesville: University Press of Florida.

———. 2000. "The Lower Creeks: Origins and Early History." In *Indians of the Greater Southeast: Historical Archaeology and Ethnohistory*, edited by B. G. McEwan, 265–298. Gainesville: University Press of Florida.

———. 2004. "Guale." In *Handbook of North American Indians, Vol. 14: Southeast*, edited by R. Fogelson and W. Sturtevant, 238–244. Washington, D.C.: Smithsonian Institution Press.

Wright, John L., Jr. 1986. *Creeks and Seminoles*. Lincoln: University of Nebraska Press.

4

The Reconstruction of Identity

A Case Study from Chachapoya, Peru

KENNETH C. NYSTROM

Despite its central position within social science research, ethnicity is a complex topic best addressed using multiple lines of evidence and flexibility in the operational definition that one assumes. Although often modeled based upon the European experience or linked to industrialization and capitalist societies (Bentley 1987), studies of ethnic identity are further advanced when considered in nontraditional contexts such as those involving precolonial state imperialism in non-Western contexts. The north coast of Peru presents such a context in which indigenous Amerindian populations were embroiled in centuries of state "cycling" and ethnic group definition and reformulation. Of particular interest to general theories of ethnicity is the relationship between imperialist expansion policies and the adaptive responses of those subjugated.

In this chapter, bioarchaeological methods are used to gain a better understanding of the relationship between pre–transitionary period biological and sociopolitical diversity and subsequent ethnogenetic processes due to major transformations in local and global political structures. Specifically, I consider patterns of pre-Inka conquest sociopolitical organization in a poorly known region of the northern Peruvian highlands (Chachapoya) in reference to subsequent presumed ethnogenesis that resulted as a sequella of Inka imperialism (Bauer 1992). Ethnohistoric and archaeological data suggest that this region was socially, and perhaps by extension ethnically, diverse. I am primarily interested in understanding the biological correlates of this observed level of diversity and what this might imply about the ethnogenetic process and the nature of ethnic groups more generally.

Population genetics models are used to quantify and interpret genetic differentiation and levels of gene flow among three samples (Kuelap, Laguna Huayabamba, and Laguna de los Cóndores) in the Chachapoya region during the Late Chachapoya period (AD 1100–1470) immediately preceding Inka conquest (ca. AD 1470). Operationally, I address two issues.

First, I compare the intraregional distributions of material markers of identity and population structure. This analysis is then related to the classic debate in anthropology about the correlation of language, biology, and culture, here manifest in archaeologically derived data sets of mortuary and architectural design elements.

This aspect of the current chapter also relates to the extensive literature comparing contemporary patterns of DNA diversity to some index of social or community organization such as ethnographic or historical data (Sokal et al. 1993; Tagarelli et al. 2005), linguistic variation (Barbujani et al. 1992; Barbujani et al. 1994; Barbujani and Sokal 1991; Bulayeva et al. 2003), religious affiliation (Barbujani et al. 1992; Smith et al. 1990), "sociopolitical attitudes" (Barbujani et al. 1992), or surnames (Madrigal et al. 2001). While these analyses often illustrate correspondences between social and biological variation in a synchronic sense, social identity is frequently conceived as a static, essentialist category. In this chapter, use of bioarchaeological data implements consideration of patterns of cultural and biological co-variation immediately preceding an ethnogenetic event, thus evaluating whether the essentialist elevation of these forms of ethnic definition is appropriate. In other words, are modern uses of ethnic labels assuming what should in fact be tested?

Second, regional levels of genetic differentiation and patterns of extralocal gene flow are presented in reference to ethnogenetic processes related to the imposition of an imperial administration and the dichotomy between the Inka and the Chachapoya this established. In consideration of the reformulation of Chachapoya identity postconquest, this aspect of the chapter better informs general theoretical perspectives on how and why ethnic groups form (see also Stojanowski, this volume; and Sutter, this volume). Did the Inka simply collapse closely related populations administratively for efficiency, or did this political structure in some way act as a catalyst for resulting Chachapoya ethnogenesis?

I begin with a brief discussion of ethnic theory as particularly incorporated into the anthropological literature. I then present contextual information on the Chachapoya region based on available archaeological and ethnohistorical data sets. I use these data to generate hypothesized correlations, which are tested using population genetic analyses based on craniometric variability. Results are interpreted in reference to Chachapoya ethnogenesis, definitions and conceptualizations of ethnic identity, and Inka imperialist modes of expansion. Therefore I not only consider how the inhabitants of Chachapoya interacted with one another but also reconstruct imperial behavior and decision-making regarding the incorporation and consolidation of conquered groups into the Inka empire.

Ethnic Identity

Early research on ethnic groups considered the following two primary issues: (1) objective versus subjective criteria for ethnic group definition and (2) whether ethnogenesis was the result of an innate quality of people (primordialism) or the result of sociopolitical and economic interaction patterns (instrumentalism). Raoul Naroll (1968: 72) proposed that the most parsimonious strategy for defining ethnic groups would be to use a concept that is "neither derived from our own culture, nor from that of any other particular culture." On the other hand, Michael Moerman (1968) advocated utilizing the subjective classification scheme of the people under investigation. For Moerman (1968: 156), the trait lists created to explain and capture the differences between ethnic groups were irrelevant to native concepts. This debate continues to resonate today as archaeologists attempt to decipher meaning and intent from material artifacts (e.g., Sackett 1982, 1985; Wiessner 1983, 1985). Aspects of this issue may also be seen in the variety of contexts, such as mortuary (Beck 1995; Larsson 1989) versus domestic (Aldenderfer and Stanish 1993; Santley et al. 1987) contexts or household versus site (Lightfoot et al. 1998) contexts, as well as in material artifacts, including textiles (Oakland Rodman 1992), ceramics (MacEachern 1992), and architecture (Hegmon 1994), which are considered to reflect ethnic identity.

The other issue that has structured ethnic theory debate is more fundamental and is concerned with ethnogenesis or how ethnic groups form (see also Stojanowski, this volume; Sutter, this volume; and Klaus and Tam Chang, this volume). Some scholars view ethnicity as resulting from a primordial, innate quality inherent to social aggregates (Geertz 1963; Issacs 1974; Keyes 1976; Shils 1957). According to primordialists, ethnic feelings are tied to the unchanging circumstances surrounding birth and growth (Issacs 1974; Shils 1957) and therefore do not account for situational shifts in identity (e.g., Nagata 1974), the fluid nature of ethnic boundaries; nor do they consider historical or contemporaneous sociopolitical contexts (Jones 1997: 70–71). This model has been criticized because of its static and atavistic conception of ethnicity (Jones 1997: 68). Alternatively, instrumentalism views ethnicity as a relatively recent sociological invention related to industrialization and capitalist economies (Bentley 1987). While the instrumental approach is more flexible and amenable for comparative analyses, it also has flaws that stem from its reductionist and ahistorical perspective (Eriksen 1991; Jones 1997), reducing ethnicity to observable behavioral regularities. The demonstration of ethnic boundaries based upon a particular form of ethnic signifier is insufficient and must also include a consideration of context. Ethnicity is a product of social context and interaction; therefore different contexts and scales of analyses (e.g., household, neighborhood, or nation) may reflect different ethnicities (Lightfoot et al. 1998). Instrumental theory is ahistorical in that it is primarily concerned with accounting for interpersonal interaction and draws attention away from the wider social and historical context (Eriksen 1991: 128). Hence ethnic groups defined as such are difficult to distinguish from other social categories or special interest groups (Jones 1997: 79). Instrumentalist concentration on sociopolitical and economic components of ethnicity also neglects the cultural and psychological aspects of ethnicity (Jones 1997: 78). There are undeniable components of ethnic group affiliation that *cannot* be reduced to sociopolitical or economic considerations. While an instrumental approach facilitates comparison because of its concentration on sociopolitical and economic factors of ethnicity, it neglects aspects of ethnicity that may be important to the agent (Eriksen 1991: 130).

The subjective and instrumental nature of social groups gained ascendancy with Fredrik Barth (1969). According to Barth (1969), traditional criteria utilized to define ethnic groups do not address boundary maintenance. Ethnogenesis follows from the isolation based upon biological, cultural, and linguistic differences. Creation of any sort of group identity in relation to "the other" relies upon perception of commonalities and differences, resulting in their reification and material objectification (Bentley 1987). On the other hand, Barth (1969) emphasized the self-referential nature of ethnic group membership, which limits the archaeological possibilities. Nonetheless, Barth (1969) is a relatively common starting point for archaeological and bioarchaeological discussions of ethnicity.

Ethnic nepotism (van den Berghe 1978) and practice theory (Bentley 1987; Jones 1997) have attempted to span this perceived primordial-instrumental dichotomy. Pierre van den Berghe's (1978) ethnic nepotism is grounded in sociobiology and views material culture as a means of extending biological relatedness. Ethnic feelings are manifested within sociopolitical, economic, and ecological settings based upon a biological disposition toward kin. In defense of ethnic nepotism, Frank Salter (2001) recognizes that, while this conception of ethnicity leads to the development of static boundaries, it allows for individual choice and therefore contestation and alteration of identity (Goetze and James 2001). A telling criticism, however, is that ethnic groups may often include fictive kin, or they may be based on a common ancestor of mythological origin, which would negate any genetic determinism that would be afforded to kin (Jones 1997).

Based in Pierre Bourdieu's (1977) theory of *habitus*, the practice theory of ethnicity contends that ethnic boundaries exist because of a set of shared dispositions, created within particular social and material environments. The theory integrates the psychological aspects of ethnicity as championed by the primordialist theory as well as the focus upon the dynamic contextual nature of ethnic identity as proposed by instrumental theory. These shared dispositions engender "feelings of identification among people similarly endowed" and are subsequently "consciously appropriated and given form through existing symbolic resources" (Bentley 1991: 173).

An individual's social identity is a dynamic amalgam of material expression and self-definition, and it is only possible for us to examine proxies of identity. Artifacts are generally considered to be the material manifestation of a commonly held group identity, but the level of analytical resolution (such as intrasite, intraregional, or interregional) and determination of context (such as mortuary or domestic) will influence the reconstruction of identity. Written documents, in contrast, presumably capture aspects of the self-referential nature of ethnic groups. Of course, such documents need to be evaluated based upon the sociopolitical environment in which they were produced, the nature and position of the authors, and, if known, the intent and purpose behind the document. Biological data such as phenotypic variance can be utilized to investigate the intersection between archaeological and written data in the reconstruction of identity.

The Chachapoya: An Introduction

The people referred to ethnohistorically as the "Chachapoya" occupied a region in northern Peru bounded by the crest of the Central Cordillera of the Andes to the west and the Río Huallaga to the east (figure 4.1).[1] Compared to the coast and central highlands of Peru, relatively little archaeological and bioarchaeological research has been conducted in the region (Bandelier 1907; Bracamonte 2002; Church 1994, 1996, 1997; Guillén 1998, 2003; Kauffmann Doig 1980, 2001, 2002; Kauffmann Doig et al. 1989; Langlois 1934, 1940a, 1940b; Morales 2002a, 2002b; Morales et al. 2002; Muscutt 1998; Narváez Vargas 1988, 1996a, 1996b; Nystrom 2004, 2005a, 2005b, 2006; Reichlen and Reichlen 1950; Ruiz Estrada 1969; Schjellerup 1980/1981,

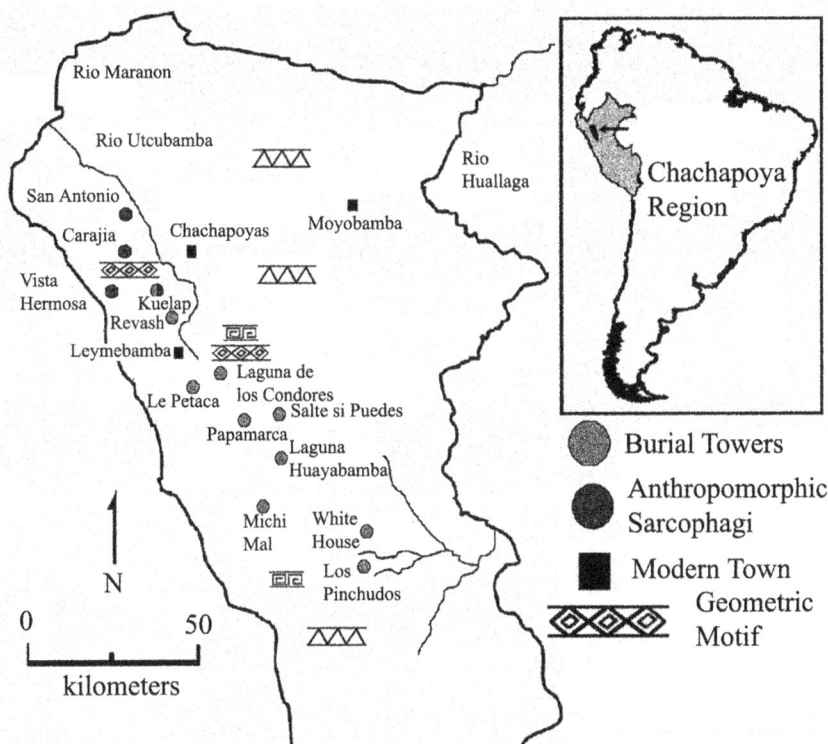

Figure 4.1. Map of the Chachapoya region of northern Peru with key archaeological sites indicated. The general north-south division in mortuary traditions and architectural geometric motifs is also represented.

1984, 1997; Urton 2001; von Hagen 2002). Based upon radiocarbon dates and excavated ceramic material, Inge Schjellerup (1997) developed the following chronological sequence: the Early Chachapoya period began in the first millennium AD, the Middle Chachapoya period was contemporaneous with late Cajamarca III and Wari style ceramics (ca. AD 600–1000), and the Late Chachapoya period (AD 1100–1470) immediately preceded the Late Horizon (ca. AD 1470–1532). In this chapter, I consider only the last of these periods.

Current interpretation of the archaeological record suggests that the Late Chachapoya period (AD 1100–1470) was a time of increased population growth, increased settlement density, and burgeoning monument construction throughout the region (Schjellerup 1997). At least two Late Chachapoya sites, Kuelap in the north and Papamarca in the Timtambo Valley in the south, appear to have been sizable (figure 4.1). Both sites have an estimated four hundred structures (Narváez Vargas 1988; Schjellerup 1997), though the large outer walls at Kuelap are not in evidence at Papamarca. While the exact sociopolitical nature and significance of these two sites remains conjectural, they were undoubtedly important centers. For example, Kuelap, constructed between AD 800 and 1100, is a massive walled site covering 450 hectares with a main walled central section measuring 582 meters long and 111 meters wide (Narváez Vargas 1988, 1996a). L. Alfredo Narváez Vargas (1988, 1996b) suggests that Kuelap was affiliated with the Chillaos ethnic group, the most important group in the area before the arrival of the Inka.

Given the relative paucity of archaeological data, it is difficult to describe sociopolitical organization during the Late Chachapoya period, but it has been characterized as a collection of semi-independent groups that united only when faced with a common adversary, the Inka (Schjellerup 1997: 67). Waldemar Espinoza Soriano (1967) noted as many as 22 separate ethnic groups in the Chachapoya region prior to Inka conquest, although his investigation was based on ethnohistorical documents and probably does not accurately reflect pre-Inka sociopolitical organization. While there appear to be regionwide similarities reflected in architecture, symbolic language, and ceramic traditions, there is some evidence for intraregional social differences as well. For example, Peter Lerche (1995) analyzed the pattern of architectural design motifs and concluded that a zigzag motif is common throughout the Chachapoya highlands, whereas rhomboid design elements are more commonly found north of the modern town of Leymebamba and interlocking spirals design elements are more common south of Leymebamba (figure 4.1). In addition, Adriana von Hagen (2002) considered the geographic distribution of mortuary structures and noted similar geographic patterning (figure 4.1), also roughly dividing the region into a northern and southern half at Leymebamba. Anthropomorphic sarcophagi, the largest of which has been described as containing a mummy (Kauffmann Doig et al. 1989: 24), are more prevalent in the northern part of the Chachapoya highlands (figure 4.2a), while burial towers or *chullpas*, although varying in terms of form (round, semicircular, rectangular, and square), are more commonly found in the south (figure 4.2b). Therefore there are several material indicators of heterogenous cultural variability in the Chachapoya region before Inka conquest.

Inka conquest initiated a number of changes in the Chachapoya region that are visible archaeologically. The Inka established an administrative center at Cochabamba just west of Papamarca, likely eclipsing the politi-

Figure 4.2. Examples of (*a*) anthropomorphic sarcophagi and (*b*) burial towers or *chullpas* found in Chachapoya (photos by Rob Dover).

cal and economic power of the latter (Schjellerup 1997). Ethnohistorical records state that Cochabamba was the seat of an *apo* (provincial governor), and the site includes many features that clearly mark it as being Inka in origin (Schjellerup 1997), including compounds with Inka *kancha*-style construction. The capacity of the storage units at Cochabamba was limited (Schjellerup 1984), and it may have been a military site rather than serving as a collection and redistribution center (Schjellerup 1997). A recently discovered site, Inka Llacta (Schjellerup et al. 2003) in the northeast periphery of the Chachapoya region, may have served as a secondary- or tertiary-level administrative center, facilitating the extraction of local resources.

Bioarchaeological Expectations for Chachapoya Ethnogenesis

This chapter considers two related issues. First, do ethnohistoric, archaeological, and biological data demonstrate corresponding levels of diversity; and if so, do the patterns of variation overlap? In other words, are the same mechanisms of ethnic diversity reflected in each form of data? Second, what does the aggregate pattern of biosocial variation imply about Inka imperialism and the nature of Chachapoya identity formation? Did the Inka ignore internal biological differences in their administrative structure? If so, this suggests that Chachapoya identity formation occurred without any substantial biological basis, contrary to primordialist expectations, and reflects a social adaptive process in accordance with an instrumentalist theory of ethnicity.

Both ethnohistoric and archaeological data suggest that the Chachapoya region during the late precontact period was ethnically diverse. If there is correspondence among these data sets, then measures of genetic diversity should also be high. This is a simple variable to measure. Unfortunately, the ethnohistoric data are poor and not useful for hypothesis testing about patterns of covariation. Espinoza Soriano (1967) notes that 22 ethnic groups were present in the region, but we have few useful details beyond this. Archaeological data, however, are more useful for model testing. In particular, two of the samples included in this analysis, Laguna Huayabamba and Laguna de los Cóndores, are located south of the presumed material boundary near Leymebamba based on mortuary structure and architectural design element differences. Therefore the pattern of biosocial correspondence can be tested by comparing intersample genetic distances across this material boundary. Most critical for this determination is the pattern of relationships for Laguna de los Cóndores. If the material boundary just north of this site affected patterns of gene flow, then two results are expected: (1) Laguna de los Cóndores is more phenotypically similar to Laguna Huayabamaba than to Kuelap; and (2) in the absence of isolation by distance throughout the region, both Laguna Huayabamba and Laguna de los Cóndores are roughly equidistant from Kuelap. A regional isolation by distance structure suggests that the hypothesized material-ethnic boundary near Leymebamba had no effect on patterns of phenotypic variation. In addition to regional isolation by distance, a homogenous Chachapoya biological population would also be indicated by: (1) low estimates of genetic divergence among samples; and (2) a similar pattern of extralocal gene flow shared throughout the region. In other words a truly homogenous biological population would have: (1) limited genetic diversity; (2) an isolation by distance population structure; and (3) shared networks of mate exchange both intra- and interregionally. Consideration of these three population parameters addresses both research questions as defined above.

Methods and Materials for Chachapoya Biodistance Analysis

For this study, 24 craniometric variables (Buikstra and Ubelaker 1994) were recorded and used to reconstruct biological relationships and patterns of genetic variability for three Late Chachapoya skeletal samples: Laguna Huayabamba, Laguna de los Cóndores, and Kuelap (figure 4.1).[2] Basic sample summary data are presented for each site in table 4.1. Preliminary statistical treatment of the data has been presented elsewhere (Nystrom 2005a, 2006). Population structure was examined using R-matrix analysis (see Relethford et al. 1997; Relethford and Blangero 1990; Schillaci and Stojanowski 2005; Steadman 1997, 2001; Stojanowski 2004, 2005, this volume), which provides

Table 4.1. Summary Data for the Three Late Chachapoya Skeletal Samples

Sample Name	n	Date	Reference
Laguna Huayabamba	17	AD 1000–1150[a]	Muscutt 2003
Laguna de los Cóndores	151	AD 1100–1420[b]	Guillén 2003
Kuelap	78	AD 1350–ca. 1470[c]	Reichlen and Reichlen 1950

[a] The radiocarbon date reported by Muscutt (2003) was from a textile sample from one of the mummy bundles found in the same cave as the crania examined for this research.
[b] The radiocarbon date reported by Guillén (2003) was derived from secondarily reconstructed bone bundles found in the same burial towers as the crania examined for this research.
[c] This is the date range originally reported by the Reichlens (1950) for the Revash cultural period, to which the crania examined for this research have been attributed.

the following three parameters of interest: (1) interpopulation genetic distances; (2) aggregate genetic differentiation among samples (F_{ST}); and (3) estimates of extralocal gene flow based on residual variance comparisons (Relethford-Blangero analysis) (Relethford and Blangero 1990; Relethford et al. 1997; Steadman 2001). These parameters allow evaluation of isolation by distance among Chachapoya populations, the overall level of phenotypic diversity present in the Chachapoya highlands immediately preceding Inka conquest, and differential mate exchange patterns with populations outside the Chachapoya heartland. Heritability was set at 0.55 (Konigsberg and Ousley 1995; but see Carson 2006), and effective population sizes were estimated based on archaeological data. Other methods of estimating effective populations did not significantly influence the results.

Results of Chachapoya Biodistance Analysis

Genetic distances, F_{ST}, and Relethford-Blangero residuals are presented in table 4.2. Interpopulation biological distances indicate that an isolation by distance structure was evident during the Late Chachapoya period. The largest genetic distance is between Kuelap, the most northern sample, and Laguna Huayabamba, the most southern sample. Laguna de los Cóndores occupies an intermediate position in terms of both geography and biodistance. With only three data points, a matrix correlation analysis (see Sutter,

Table 4.2. Minimum Genetic Distances and Residual Variance Analysis for Late Chachapoya Populations (heritability = 0.55)

GENETIC DISTANCES

	Huayabamba	Cóndores	Kuelap
Huayabamba	0.000		
Cóndores	0.164	0.000	
Kuelap	0.447	0.181	0.000

RESIDUAL VARIANCE ANALYSIS

	r_{ii}	$\check{v}_i$	$E_{(\check{v}i)}$	$\check{v}_i - E_{(\check{v}i)}$
Huayabamba	0.243	0.931	0.724	0.207
Cóndores	0.111	1.006	0.851	0.154
Kuelap	0.035	0.821	0.924	-0.102
$F_{ST} = 0.090$, $se = 0.013$				

r_{ii}: regional phenotypic distances to centroid.
$\check{v}_i$: observed mean variance.
$E_{(\check{v}i)}$: expected mean variance.
$\check{v}_i - E_{(\check{v}i)}$: residual variance.

this volume) would not be very informative, but the results are surely meaningful. The aggregate level of genetic differentiation (F_{ST}) among populations is estimated as 0.090, which is moderately large. Given the pattern of population relationships and magnitude of F_{ST} the Chachapoya were weakly united by long-range migration but were not a homogenous biological population. Relethford-Blangero residual analysis indicates that the population at Kuelap was interacting less than expected with extralocal populations. Both Laguna Huayabamba and Laguna de los Cóndores populations returned positive residuals, indicating more extralocal gene flow than average. Interestingly, residuals become less positive as one moves from south to north and again indicate heterogeneity in practice and experience by different populations inhabiting the Chachapoya region.

Discussion

Results presented in this chapter indicate that the late preconquest Chachapoya exhibited an isolation by distance population structure, reflecting limited long-range gene flow throughout the region that nonetheless did not create a homogenous biological population. In addition, there was no correspondence between the distribution of presumed material manifestations of subgroup identity (such as mortuary sites and geometric motifs) and population structure. The presence of a symbolic boundary of ethnic identity around the Leymebamba region did not affect the pattern of hypothesized within- and between-region genetic distances. Despite potential ethnic group differentiation, as defined by archaeological material, the phenotypic data indicate that, if these cultural barriers did exist, they did not significantly influence patterns of gene flow among populations.

Regional measures of genetic differentiation ($F_{ST} = 0.090$) are difficult to evaluate in reference to pre-Inka ethnic group composition. Current archaeological interpretation suggests that the region was inhabited by a semi-independent confederation of groups that only united in the face of their common enemy, the Inka. Measures of regional genetic differentiation are high and support this archaeological interpretation. Because we do not have adequate skeletal samples from either earlier or later periods, however, it is impossible to evaluate this figure as evidence for consolidation of different Chachapoya groups (ethnogenesis?) in the face of Inka conquest (see Stojanowski, this volume).

Is this amount of genetic differentiation what one would expect for what is assumed to be a single "people"? While it is greater than what is typically

Table 4.3. Comparative F_{ST} Estimates for Amerindian Populations

Population	F_{ST}	Distance (km)	Time (Yrs)	Source
Florida Late Precontact	0.008	100	300	Stojanowski 2004
Illinois Woodland	0.004	150	500	Steadman 1997, 2001
Illinois Mississippian	0.010	150	500	Steadman 1997, 2001
Ohio Late Archaic	0.039	150	650	Tatarek and Sciulli 2000
Ohio Late Prehistoric	0.078	300	650	Tatarek and Sciulli 2000
Chachapoya	0.090	75	300	this study

expected for small tribal populations (Jorde 1980) in general, the F_{ST} value for the Late Chachapoya populations is in the upper range represented by other Amerindian groups (table 4.3). While F_{ST} values are not directly comparable between regions, we can broadly compare these results to previous publications that also used phenotypic data in archaeological populations to obtain a rough evaluation of how genetically variable the Chachapoya region was during this period. In general, it appears that the Late Chachapoya groups, even though sampled from a geographically more restricted area (approximately 75 km) than comparative estimates and representing a shorter temporal span in terms of archaeological duration of sites (approximately 300 years), exhibit higher levels of genetic differentiation. In other words, if one normalizes the data for two variables that increase estimates of regional genetic diversity in archaeological samples (distance and time sampled), the Chachapoya are more heterogeneous than expected. This result is fairly robust and confirms the ethnohistoric and archaeological data that also suggest commensurate social variability.

The pattern of Relethford-Blangero residuals also attests that the Chachapoya were not a single homogenous biological population. Chachapoya populations did not share a similar interaction pattern with other regions, at least as represented by biological diversity estimates for these three samples. Populations in the southern highland region had greater interaction with extralocal populations than expected, whereas the sample from Kuelap in the north indicated less than average interaction. The patterns of genetic distances and Relethford-Blangero residuals indicate clinal changes in biological variation from two independent measures, which is intriguing and, at face value, suggests a south-to-north flow of genes throughout the Chachapoya highlands. While this pattern may be the result of some type of directional exogamous marriage system, there is little evidence for this type of movement; in fact, existing evidence suggests connections to the west, in the form of Cajamarca ceramics that have been found throughout the region (Muscutt 2003; Schjellerup 1997; von Hagen 2002). In addition, extensive work at Laguna de los Cóndores produced evidence for Amazonian trade connections for this centrally located sample. Materials recovered from Laguna de los Cóndores, including a parrot feather headdress, the tanned remains of two small felines (*Felis wiedii* and *Felis tigrina*) native to tropical environments, and iconographic elements, point to a strong connection with the Amazonia Basin (von Hagen 2002). Warren Church (1996: 7), in his examination of Manachaqui Cave, suggests that the tropical montane societies may have acted as "primary purveyors and intermediary conveyors" in economic relations that connected other Andean regions with eastern Amazonia regions. We are unable to evaluate this pattern, however, because Laguna Huayabamba has not been as intensively excavated. It is possible that a trade connection with the Amazonia Basin may explain the positive residual obtained for the Laguna de los Cóndores, assuming commensurate movement of genes and materials.

In addition, although the evidence is based on only three skeletal samples, it appears the Inka did not collapse a single "people" into an administrative unit; rather, they glossed over a certain degree of intraregional subgroup differences marked by material culture. Additionally, they incorporated phenotypically variable subgroups that, while united by some internal gene flow, were experiencing different levels of external gene flow. Interestingly, the Spanish perceived the Chachapoya as distinct from other Andean populations. According to Pedro de Cieza de León (1998 [1553]: 99), the Chachapoya were "the whitest and most attractive I have seen anywhere." Although this implies nothing about internal genetic variability, it does suggest the possibility that Inka-Chachapoya interactions may have been based in part on perceived phenotypic differences.

These results also provide insight into Inka imperial decision-making practices and what they may have used as their primary criteria in the creation of their administrative units. I would argue that we (at least implicitly) have assumed that these units were formed by the Inka based upon their recognition of material culture differences. Indeed, the Inka commonly required conquered groups to maintain regionally distinctive dress as a means of identification. The results from this research support this general contention, as the Inka appear to have collapsed regional subgroups based upon material similarities that were only loosely connected by internal gene flow.

In this sense, it appears that Inka conquest served as a catalyst for ethnogenesis (the formation of regional-level group identity). Imperially imposed administrative units were ostensibly predicated upon some external and/ or internal group definition (Jenkins 2003); and while the groups that inhabited the Chachapoya region may have recognized internal differences, it seems that the Inka considered them to be a single group and treated them as such. This research indicates that Chachapoya ethnogenesis occurred because of Inka imperialist policy and the dialectical structure of economic and political relationships that it created. This interpretation is consistent with the theory that social identity is created during the interaction between groups (Barth 1969) and the perception of commonalities and differences, resulting in reification and objectification (Bentley 1987). The form that this objectification takes is context specific and dynamic, and there is not a trait-list that will unequivocally delimit one group from another. While material manifestations are powerful markers of group difference, they are only a single aspect of group identity. Similarly, group identity cannot be adequately defined based on biodistance measures or trait frequency. By considering and articulating material and biological data, however, it is possible to gain a deeper insight into the processes of identity formation.

Conclusion

Analysis of biological data indicates that during the Late Chachapoya period the region was inhabited by phenotypically variable populations connected by limited internal gene flow. Furthermore, presumed material markers of intraregional identity such as architectural and mortuary design elements were not significant barriers to gene flow within the Chachapoya region. We can say, therefore, that the region inhabited by the people known as the Chachapoya by the Inka was not biologically or materially homogenous in the period immediately preceding Inka conquest. This suggests that the label "Chachapoya" (just as with racial typological labels) may have had little biological meaning, at least during the precontact period. Unfortunately, we currently lack data to evaluate whether biological homogenization occurred after Inka conquest, a research model similar to that documented by Stojanowski (this volume). Spanish sources from the postcontact *entrada* chronicles, however, at least imply that this homogenization occurred, bolstering the importance of situational contexts in ethnic group formation.

Acknowledgments

This research was funded in part by National Science Foundation Doctoral Improvement Grant 0242941, the Latin American and Iberian Institute, and the Graduate Students and Professional Association of the University of New Mexico. I would to thank Claudia Grimaldo, Sonia Guillén, Jesus Briceño, and Keith Muscutt for facilitating access to the skeletal samples. Thanks also to John Relethford for providing access to the RMET 5.0 program and for answering procedural questions. All statistical errors remain solely the responsibility of the author.

Notes

1. Ethnohistoric sources for the Chachapoya are limited and include writings by Padre Blas Valera, Garcilaso de la Vega, Pedro de Cieza de León, Sarmiento de Gamboa, and Guaman Poma de Ayala (see Adorno 2000; Schjellerup 1997: 61).

2. The following craniometric variables were used in this study: Maximum Cranial Length, Maximum Cranial Breadth, Bi-Zygomatic Breadth, Basion-Bregma Height, Cranial Base Length, Basion-Prosthion Length, Maximum Aveolar Breadth, Maximum Aveolar Length, Bi-Auricular Breadth, Upper Facial Height, Upper Facial Breadth, Nasal Height, Nasal Breadth, Orbital Height, Orbital Breadth, Bi-Orbital Breadth, Inter-Orbital Breadth, Frontal Chord, Parietal Chord, Occipital Chord, Foramen Magnum Length, Foramen Magnum Breadth, and Mastoid Length.

References Cited

Adorno, Rolena. 2000. *Guaman Poma: Writing and Resistance in Colonial Peru*. Austin: University of Texas Press.

Aldenderfer, Mark S., and Charles Stanish. 1993. "Domestic Architecture, Household Archaeology, and the Past in the South-Central Andes." In *Domestic Architecture, Ethnicity, and Complementarity in the South-Central Andes*, edited by M. S. Aldenderfer, 1–12. Iowa City: University of Iowa Press.

Bandelier, Adolf. 1907. *The Indians and Aboriginal Ruins near Chachapoyas in Northern Peru*. Historical Records and Studies. New York: United States Catholic Historical Society.

Barbujani, Guido, Ivane S. Nasidze, and Gerard N. Whitehead. 1994. "Genetic Diversity in the Caucasus." *Human Biology* 66: 639–668.

Barbujani, Guido, and Robert R. Sokal. 1991. "Genetic Population Structure of Italy, II: Physical and Cultural Barriers to Gene Flow." *American Journal of Human Genetics* 48: 398–411.

Barbujani, Guido, Paolo Vian, and Luigi Fabbris. 1992. "Cultural Barriers Associated with Large Gene Frequency Differences among Italian Populations." *Human Biology* 64: 479–495.

Barth, Fredrik. 1969. "Introduction." In *Ethnic Groups and Boundaries: The Social Organization of Culture Difference*, edited by F. Barth, 9–38. Boston: Little, Brown and Company.

Bauer, Brian S. 1992. *The Development of the Inca State*. Austin: University of Texas Press.

Beck, Lane A. 1995. "Regional Cults and Ethnic Boundaries in 'Southern Hopewell.'" In *Regional Approaches to Mortuary Analysis*, edited by L. A. Beck, 167–187. New York: Plenum Press.

Bentley, G. Carter. 1987. "Ethnicity and Practice." *Comparative Studies in Society and History* 29: 24–55.

———. 1991. "Response to Yelvington." *Comparative Studies in Society and History* 33: 169–175.

Bourdieu, Pierre. 1977. *Outline of a Theory of Practice*. Cambridge: Cambridge University Press.

Bracamonte, G. Florencia. 2002. "Los Pinchudos: Un estudio preliminar de su población." *Sian* 8: 14–15.

Buikstra, Jane E., and Douglas H. Ubelaker, editors. 1994. *Standards for Data Collection from Human Skeletal Remains*. Arkansas Archeological Survey Research Series No. 44. Fayetteville: Arkansas Archeological Survey.

Bulayeva, Kazima, Lynn B. Jorde, Christopher Ostler, Scott Watkins, Oleg Bulayev, and Henry Harpending. 2003. "Genetics and Population History of Caucasus Populations." *Human Biology* 75: 837–853.

Carson, E. Anne. 2006. "Maximum Likelihood Estimation of Human Craniometric Heritabilities." *American Journal of Physical Anthropology* 131: 169–180.

Church, Warren B. 1994. "Early Occupations at Gran Pajatén, Peru." *Andean Past* 4: 281–318.

———. 1996. "Prehistoric Cultural Development and Interregional Interaction in the Tropical Montane Forests of Perú." Doctoral dissertation, Yale University, Department of Anthropology.

———. 1997. "Mas allá del Gran Pajatén: Conservando el paisaje prehispánico Pataz-Abiseo." *Revista del Museo de Arqueología, Antropología e Historia* 7: 205–248.

Cieza de León, Pedro de. 1998 [1553]. *The Discovery and Conquest of Peru: Chronicles of the New World Encounter*. Edited and translated by A. P. Cook and N. D. Cook. Durham and London: Duke University Press.

Eriksen, Thomas H. 1991. "The Cultural Contexts of Ethnic Differences." *Man* 26: 127–144.

Espinoza Soriano, Waldemar. 1967. "Los señoríos étnicos de Chachapoyas y la alianza hispano-chacha: Visitas, informaciones y memoriales inéditos de 1572–1574." *Revista Histórica* 30: 224–283.

Geertz, Clifford. 1963. "The Integrative Revolution: Primordial Sentiments and Civil Politics in the New States." In *Old Societies and New States: The Quest for Modernity in Asia and Africa*, edited by C. Geertz, 105–157. London: Free Press of Glencoe.

Goetze, David, and Patrick James. 2001. "What Can Evolutionary Theory Say about Ethnic Phenomena?" In *Evolutionary Theory and Ethnic Conflict*, edited by P. James and D. Goetze, 3–18. London: Praeger.

Guillén, Sonia E. 1998. "Laguna de los Cóndores: Donde viven los muertos." *Bien Venida* 6: 43–48.

———. 2003. "Keeping Ancestors Alive: The Mummies from Laguna de los Cóndores, Amazonas, Perú." In *Proceedings of the 4th World Congress on Mummy Studies, Nuuk, Greenland, September 4–10, 2001*, edited by N. Lynnerup, C. Andreasen, and J. Berglund, 162–164. Copenhagen: Greenland National Museum and Archives and Danish Polar Center.

Hegmon, Michelle. 1994. "Boundary-Making Strategies in Early Pueblo Societies: Style and Architecture in the Kayenta and Mesa Verde Regions." In *The Ancient Southwestern Community: Models and Methods for the Study of Prehistoric Social Organization*, edited by W. H. Wills and R. D. Leonard, 171–188. Albuquerque: University of New Mexico Press.

Issacs, Harold. 1974. "Basic Group Identity: The Idols of the Tribe." *Ethnicity* 1: 15–41.

Jenkins, Richard. 2003. "Rethinking Ethnicity: Identity, Categorization, and Power." In *Race and Ethnicity: Comparative and Theoretical Approaches*, edited by J. Stone and R. Dennis, 59–71. Oxford: Blackwell Publishers.

Jones, Siân. 1997. *The Archaeology of Ethnicity: Constructing Identities in the Past and Present*. London and New York: Routledge.

Jorde, Lynn B. 1980. "The Genetic Structure of Subdivided Human Populations." In *Current Developments in Anthropological Genetics*, edited by J. H. Mielke and M. H. Crawford, 135–208. New York: Plenum Press.

Kauffmann Doig, Federico. 1980. "'Los Pinchudos': Exploración de ruinas intactas en la selva." *Boletín de Lima* 7: 26–31.

———. 2001. "Los Sarcófagos de Carajía: A 15 años de un descubrimiento sensacional." *Arkinka* 65: 76–91.

———. 2002. *Historia y arte del Perú antiguo*. Vol. 4. Lima, Peru: PEISA.

Kauffmann Doig, Federico, Miriam Salazar, Daniel Morales, Iain Mackay, and Oscar Sacay. 1989. "Andes Amazónicos: Sitios intervenidos por la expedición Antisuyo/86." *Arqueológicas* 20: 6–57. Lima: Museo Nacional de Antropología y Arqueología.

Keyes, Charles F. 1976. "Towards a New Formulation of the Concept of Ethnic Group." *Ethnicity* 3: 202–213.

Konigsberg, Lyle W., and Stephen D. Ousley. 1995. "Multivariate Quantitative Genetics of Anthropometric Traits from the Boas Data." *Human Biology* 67: 481–498.

Langlois, Louis. 1934. "Las ruinas de Cuelap." *Boletín de la Sociedad Geográfica de Lima* 51: 20–34.

———. 1940a. "Utcubamba: Investigaciones arqueológicas en este valle del Departamento de Amazonas (Perú), Continuación." *Revista del Museo Nacional* 9: 33–72.

———. 1940b. "Utcubamba: Investigaciones arqueológicas en este valle del Departamento de Amazonas (Perú), Conclusión." *Revista del Museo Nacional* 9: 191–225.

Larsson, Lars. 1989. "Ethnicity and Traditions in Mesolithic Mortuary Practices of Southern Scandinavia." In *Archaeological Approaches to Cultural Identity*, edited by S. J. Shennan, 210–218. London: Unwin Hyman.

Lerche, Peter. 1995. *Los Chachapoya y los símbolos de su historia*. Lima: n.p.

Lightfoot, Kent G., Antoinette Martinez, and Ann M. Schiff. 1998. "Daily Practice and Material Culture in Pluralistic Social Settings: An Archaeological Study of Culture Change and Persistence from Fort Ross, California." *American Antiquity* 63: 199–222.

MacEachern, Scott. 1992. "Ethnicity and Stylistic Variation around Mayo Plata, Northern Cameroon." In *An African Commitment: Papers in Honour of Peter Lewis Shinnie*, edited by J. Sterner and N. David, 211–230. Calgary: University of Calgary Press.

Madrigal, L., B. Ware, R. Miller, G. Saenz, M. Chavez, and D. Dykes. 2001. "Ethnicity, Gene Flow, and Population Subdivision in Limón, Costa Rica." *American Journal of Physical Anthropology* 114: 99–108.

Moerman, Michael. 1968. "Being Lue: Uses and Abuses of Ethnic Identification." In *Essays on the Problem of Tribe: Proceedings of the 1967 Annual Spring Meeting of the American Ethnological Society*, edited by J. Helm, 153–169. Seattle: University of Washington Press.

Morales, G. Ricardo. 2002a. "Los Pinchudos: Al rescate de una arquitectura funeraria en emergencia, II parte." *Arkinka* 79: 86–99.

———. 2002b. "Los Pinchudos: Arquitectura funeraria en Río Abiseo, San Martín, I parte." *Arkinka* 76: 92–101.

Morales, G. Ricardo, Luis V. Alvarez, Warren B. Church, and Luis C. Tello. 2002. "Los Pinchudos: Un estudio preliminar de su población." *Sian* 8: 1–41.

Muscutt, Keith. 1998. *Warriors of the Clouds: A Lost Civilization in the Upper Amazon of Peru*. Albuquerque: University of New Mexico Press.

———. 2003. Unpublished History Channel Site Report on Vira Vira, Peru.

Nagata, Judith A. 1974. "What Is Malay?: Situational Selection of Ethnic Identity in a Plural Society." *American Ethnologist* 1: 331–350.

Naroll, Raoul. 1968. "Who the Lue Are." In *Essays on the Problem of Tribe: Proceedings of the 1967 Annual Spring Meeting of the American Ethnological Society*, edited by J. Helm, 72–79. Seattle: University of Washington Press.

Narváez Vargas, L. Alfredo. 1988. "Kuelap: Una ciudad fortificada en los Andes nororientales de Amazonas, Perú." In *Arquitectura y arqueología: pasado y futuro de la construcción en el Perú*, edited by V. Rangel Flores, 115–142. Chiclayo: Universidad de Chiclayo, Perú.

———. 1996a. "La fortaleza de Kuélap." *Arkinka* 12: 92–108.

———. 1996b. "La fortaleza de Kuélap." *Arkinka* 13: 90–98.

Nystrom, Kenneth C. 2004. "Trauma y identidad entre los Chachapoya." Primera Conferencia Internacional sobre el Arte, la Arqueología y la Etnohistoria de los Chachapoya, Leymebamba, Perú. *Sian* 9, no. 15: 20–21.

———. 2005a. "The Biological and Social Consequences of Inka Conquest of the Chachapoya Region of Northern Perú." Doctoral dissertation, University of New Mexico, Department of Anthropology.

———. 2005b. "Chachapoya Mummies from the Laguna Huayabamba." Proceedings of the 5th World Mummy Congress, Turin, Italy. *Journal of Biological Research* 80: 197–200.

———. 2006. "Late Chachapoya Population Structure Prior to Inka Conquest." *American Journal of Physical Anthropology* 131: 334–342.

Oakland Rodman, Amy. 1992. "Textiles and Ethnicity: Tiwanaku in San Pedro de Atacama, North Chile." *Latin American Antiquity* 3: 316–340.

Reichlen, Henry, and Paule Reichlen. 1950. "Recherches Archéologiques dans les Andes du Haut Utcubamba." *Journal de la Société des Américanistes* 39: 219–246.

Relethford, John H., and John Blangero. 1990. "Detection of Differential Gene Flow from Patterns of Quantitative Variation." *Human Biology* 62: 5–25.

Relethford, John H., Michael H. Crawford, and John Blangero. 1997. "Genetic Drift and Gene Flow in Post-Famine Ireland." *Human Biology* 69: 443–465.

Ruiz Estrada, Arturo. 1969. "Alfarería del estilo Huari en Cuelap." *Boletín del Seminario de Arqueología* 4: 60–65.

Sackett, James R. 1982. "Approaches to Style in Lithic Archaeology." *Journal of Anthropological Archaeology* 1: 59–112.

———. 1985. "Style and Ethnicity in the Kalahari: A Reply to Wiessner." *American Antiquity* 50: 154–159.

Salter, Frank. 2001. "A Defense and an Extension of Pierre van den Berghe's Theory of Ethnic Nepotism." In *Evolutionary Theory and Ethnic Conflict*, edited by P. James and D. Goetze, 39–70. London: Praeger.

Santley, Robert, Clare Yarborough, and Barbara Hall. 1987. "Enclaves, Ethnicity, and the Archaeological Record at Matacapan." In *Ethnicity and Culture: Proceedings of the Eighteenth Annual Conference of the Archaeological Association of the University of Calgary*, edited by R. Auger, M. F. Glass, S. MacEachern, and P. H. McCartney, 85–100. Calgary: University of Calgary Archaeological Association.

Schillaci, Michael A., and Christopher M. Stojanowski. 2005. "Craniometric Variation and Population History of the Prehistoric Tewa." *American Journal of Physical Anthropology* 126: 404–412.

Schjellerup, Inge. 1980/1981. "Documents on Stone and in Paper: A Preliminary Report on Cochabamba, an Inca Administrative Center." *Folk* 22/23: 299–311.

———. 1984. *Cochabamba—an Inca Administrative Centre in the Rebellious Province of Chachapoyas.* BAR International Series 210. Oxford: Archaeopress.

———. 1997. *Incas and Spaniards in the Conquest of the Chachapoyas, Archaeological and Ethnohistorical Research in the North-eastern Andes of Perú.* GOTARC, series B, Gothenburg Archaeological Theses, 7. Göteborg University.

Schjellerup, Inge, Mikael K. Sørensen, Carolina Espinoza, Victor Quipuscoa, and Victor Peña. 2003. *The Forgotten Valleys: Past and Present in the Utilization of Resources in the Ceja de Selva, Perú.* Ethnographic Monographs No. 1. Copenhagen: National Museum of Denmark.

Shils, Edward. 1957. "Primordial, Personal, Sacred and Civil Ties: Some Particular Observations on the Relationships of Sociological Research and Theory." *British Journal of Sociology* 8: 130–145.

Smith, M. T., W. R. William, J. J. McHugh, and A. H. Bittles. 1990. "Isonymic Analysis of Post-Famine Relationships in the Ards Peninsula, N. E. Ireland: Effects of Geographical and Politico-Religious Boundaries." *American Journal of Human Biology* 2: 245–254.

Sokal, Robert R., Geoffrey M. Jacquez, Neal L. Oden, Donna DiGiovanni, Anthony B. Falsetti, Elizabeth McGee, and Barbara A. Thomson. 1993. "Genetic Relationships of European Populations Reflect Their Ethnohistorical Affinities." *American Journal of Physical Anthropology* 91: 55–70.

Steadman, Dawnie W. 1997. "Population Genetic Analysis of Regional and Interregional Prehistoric Gene Flow in West-Central Illinois." Doctoral dissertation, University of Chicago, Department of Anthropology.

———. 2001. "Mississippians in Motion?: A Population Genetic Analysis of Interregional Gene Flow in West-Central Illinois." *American Journal of Physical Anthropology* 114: 61–73.

Stojanowski, Christopher M. 2004. "Population History of Native Groups in Pre- and Postcontact Spanish Florida: Aggregation, Gene Flow, and Genetic Drift on the Southeastern U.S. Atlantic Coast." *American Journal of Physical Anthropology* 123: 316–332.

———. 2005. "The Bioarchaeology of Identity in Spanish Colonial Florida: Social and Evolutionary Transformation before, during, and after Demographic Collapse." *American Anthropologist* 107: 417–431.

Tagarelli, A., A. Piro, G. Tagarelli, P. Lagonia, A. Bulo, A. Falchi, L. Varesi, G. Vona, and C. M. Calò. 2005. "Genetic Characterization of the Historical Albanian Ethnic Minority of Calabria (Southern Italy)." *Human Biology* 77: 45–60.

Tatarek, Nancy E., and Paul W. Sciulli. 2000. "Comparison of Population Structure in Ohio's Late Archaic and Late Prehistoric Periods." *American Journal of Physical Anthropology* 112: 363–376.

Urton, Gary. 2001. "A Calendrical and Demographic Tomb Text from Northern Peru." *Latin American Antiquity* 12: 127–148.

van den Berghe, Pierre L. 1978. "Race and Ethnicity: A Sociobiological Perspective." *Ethnic and Racial Studies* 1: 401–411.

von Hagen, Adriana. 2002. "Chachapoya Iconography and Society at Laguna de los Cóndores, Perú." In *Andean Archaeology Volume II: Art, Landscape, and Society*, edited by H. Silverman and W. H. Isbell, 137–155. New York: Kluwer Academic/Plenum Publishers.

Wiessner, Polly. 1983. "Style and Social Information in Kalahari San Projectile Points." *American Antiquity* 48: 253–276.

———. 1985. "Style or Isochrestic Variation?: A Reply to Sackett." *American Antiquity* 50: 160–166.

5

Post-Tiwanaku Ethnogenesis in the Coastal Moquegua Valley, Peru

RICHARD C. SUTTER

While ethnogenesis can result when unrelated outsiders co-opt a group's identity (Haley and Wilcoxon 2005), it can also occur *in situ* as individuals shed old social identities and develop new ones in the social and political vacuum created during state collapse. How, then, can we identify when culturally expressed group identity is crafted by those who are co-opting others' identities, versus instances when new identities are forged by people who, while rejecting their former identity, are constrained by history and tradition? I explore these questions in the Moquegua Valley of the South Central Andes, where the Chiribaya polity emerged during the collapse of terminal Middle Horizon (AD 750–1100) middle valley Tiwanaku colonies (figure 5.1). The Chiribaya provide us with an unprecedented opportunity to examine ethnogenesis, using dentally derived biodistance data to test two alternative explanations for Chiribaya genetic and cultural origins. When considered in light of site survey and archaeological data, my biodistance results indicate that the Chiribaya were largely descended from Tiwanaku colonists who migrated to the coast following the abandonment of their middle valley settlements sometime around AD 900. As Tiwanaku identity lost its salience in the Moquegua Valley, some of their former colonists migrated to the coast and developed new designs to express their identity through ceramics and textiles, while largely retaining previous mortuary practices, ceramic forms, house construction, and other unconsciously expressed aspects of their formerly Tiwanaku ethnic identity. I discuss these results within the context of relevant social theory regarding economic models of cooperation as they relate to ethnogenesis. I begin with a general overview of the cultural history for the Moquegua Valley to contextualize the emergence of ethnically Chiribaya peoples.

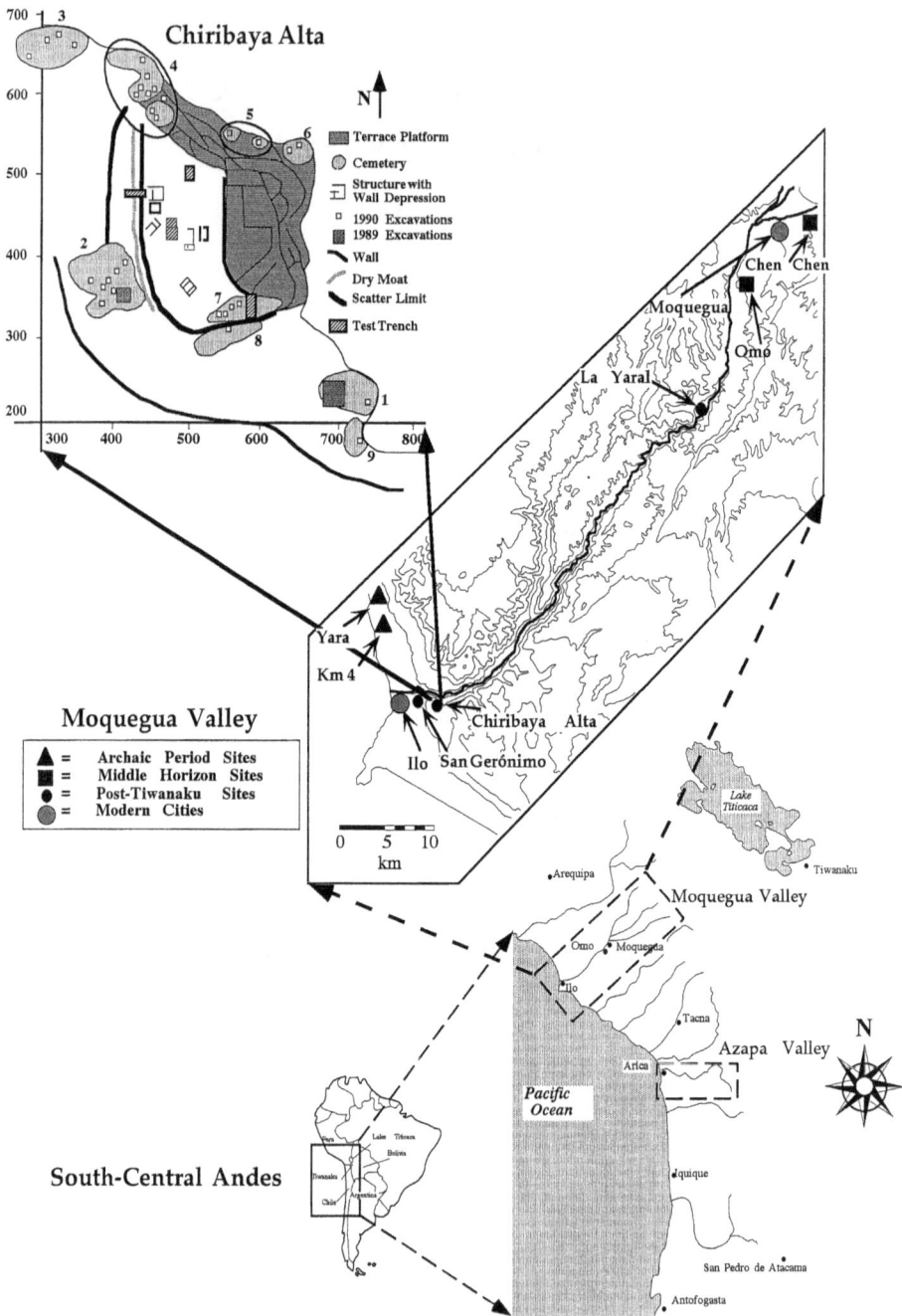

Figure 5.1. The South Central Andes, including the Moquegua Valley and the post-Tiwanaku influence site of Chiribaya Alta.

Culture History of the Moquegua Valley

Sites identified by archaeological surveys and excavations indicate that during the Archaic Period (8000–1000 BC) coastal sites in the Moquegua region were located adjacent to spring-fed *quebradas* (dry ravines) to the north and along the pampa to the south coast of the modern city of Ilo. Sites consisted of large shell mounds, lithic debris and fishing tackle scatters, residential terraces, and Chinchorro group burials. This population concentration along the coast contrasts with the paucity of Archaic sites for the middle Moquegua Valley located inland. Formative Period (1000 BC–AD 500) Huaracane sites in the middle Moquegua Valley represent the earliest agricultural settlements of the region and roughly date between 800 BC and AD 500 (Goldstein 2005). Bruce Owen (1993: 412) reports that Algodonal Early Ceramics from lower Moquegua Valley sites demonstrate similarities with ceramics associated with middle valley Huaracane sites (Owen 1993: 412, 2005: 51).

The Middle Horizon (ca. AD 500–1100) is defined by the spread of Tiwanaku V ceramics throughout the South Central Andes (see also Knudson and Blom, this volume). In the middle Moquegua Valley there is substantial evidence for numerous contemporary settlements affiliated with both the southern highland Wari and *altiplano* (high plateau) Tiwanaku states (Blom et al. 1998; Goldstein 2005; Knudson and Blom, this volume). Both Wari and Tiwanaku IV and V sites are located at altitudes ranging between 1,000 and 2,000 m.a.s.l. (meters above sea level). Paul Goldstein (2005) argues that the sunken court at Omo M10 and the presence of Tiwanaku IV and V serving wares and blackwares from ceremonial, funerary, and domestic contexts attest to permanent Tiwanaku colonies in the middle Moquegua Valley. Based upon ceramics, mortuary features, and settlement patterns, he identifies two contemporaneous ethnically Tiwanaku traditions in the middle Moquegua Valley that he refers to as the Omo and Chen Chen styles. Goldstein (2005: 158) reports radiocarbon dates that range from cal AD 785 to cal AD 1000 for the Chen Chen style, while the Omo style ranges from AD 538 to 1030 (Goldstein 2005: 152). While he argues that the slightly higher elevation settlements associated with Chen Chen–style ceramics represent Tiwanaku agricultural colonies with cultural ties to the southern Lake Titicaca region (Goldstein 2005: 223), the blackware ceramics from Omo-style settlements exhibit strong similarities to those from the Copacabana Peninsula and eastern slopes of Bolivia (Goldstein 2005: 316). Omo-style settlements are located in slightly lower elevations on bluffs farther from agricultural fields than contemporaneous Chen Chen–style settlements. Furthermore, Omo-style settlements are characterized by relatively fewer and smaller grinding stones, ephemeral cane domestic structures, and paths radiating from them, leading Goldstein (2005: 153–154) to suggest that the Omo-style Tiwanaku colonists likely represent camelid herders and traders. Among other important distinctions from the Chen Chen–style settlement was the preponderance of textile implements in both Omo-style domestic structures and female grave offerings (Goldstein 2005: 200, 254). Both mitochondrial DNA (Lewis et al. 2007) and epigenetic cranial trait comparisons (Blom et al. 1998; Knudson and Blom, this volume) of a Chen Chen mortuary sample with remains from Tiwanaku and other Lake Titicaca sites support Goldstein's assertion that Tiwanaku colonists resided in the middle valley. Similarly, strontium isotope analyses by Kelly Knudson (Knudson and Blom, this volume; Knudson and Price 2007) provide evidence that at least some of those interred at the site of Chen Chen were raised in the southern Lake Titicaca region.

Archaeologists have documented the dramatic abandonment of Tiwanaku settlements during the late Middle Horizon at both Omo- and Chen Chen–style sites, concurrent with an apparent dispersal of the colonists who previously resided there (Goldstein 2005; Owen 2005). Ryan Patrick Williams (2002) proposes that this dispersal was largely due to the progressive abandonment of Tiwanaku agricultural fields and settlements due to pressures imposed by Wari irrigation works built upstream from Tiwanaku Omo-style and Chen Chen–style colonies. Tombs were looted at Chen Chen–style sites, and the temple at Omo M10 was destroyed. Pitted rock piles from this destruction were left at Chen Chen–style sites, while the inhabitants dispersed to fortified or naturally protected sierra settlements in previously uninhabited regions of the upper tributaries of the Moquegua Valley and established a similar but new ceramic style referred to as Tumilaca (Goldstein 2005; Owen 2005). Wari colonists on and around Cerro Baúl later abandoned their settlements and left the region too. There is clear evidence of a strong degree of cultural continuity among Tumilaca and Tiwanaku Chen Chen–style ceramics, mortuary patterns, and grave furniture. Middle valley Tumilaca ceramics share similar forms and designs with previous Chen Chen–style ceramics but are more variable and poorly executed and lack the Gateway God iconography that is commonly found on earlier Tiwanaku ceramics from both the adjacent *altiplano* and middle valley colonies (Goldstein 2005: 233).

Based upon his extensive survey and excavation of stratified habitation sites, Owen (1993, 2005) asserts that the culture history for the lower Moquegua Valley region is markedly different than that for the middle valley during the Middle Horizon. Unlike the middle valley settlement pattern, Owen (1993: 58, 2005: 58) reports a lack of evidence for Tiwanaku sites on the coast. Instead, early Middle Horizon lower valley sites continue to be characterized by the Algodonal Early Ceramic tradition, which exhibits no stylistic or technical similarities with contemporaneous Tiwanaku ceramics (Owen 2005: 66). The Early Ceramic tradition of the lower valley is strikingly similar to the Huaracane traditions of the middle valley (Owen 1993: 412, 2005: 51). Calibrated radiocarbon dates for the Early Ceramic tradition range between 100 BC and AD 610 (Owen 2005: 72). There are relatively few small Early Ceramic sites, leading Owen (1993, 2005) to conclude that the lower valley was sparsely populated by fisher-agriculturalists during most of the Middle Horizon until approximately AD 900, when this population was absorbed by an overwhelming immigration of post-Tiwanaku middle valley peoples of the Ilo-Tumilaca/Cabuza (ITC) and Chiribaya traditions.

The emergence of both the coastal ITC and Chiribaya traditions has been referred to as part of a two-stage diaspora of Tiwanaku peoples (Goldstein 2005: 323; Owen 2005). It is argued that the collapse and dispersal of Tiwanaku's colonists from the middle valley also led to population movements into the lower Moquegua Valley (Owen 1993: 91–98, 2005: 65–67). Owen (1993: 94–98, 2005) has established that both Chiribaya and ITC traditions are distinct on the basis of ceramics, mortuary practices, textiles, spoons, diet, and domestic architecture, yet these two cultural traditions peacefully coexisted for centuries; Chiribaya and ITC settlements shared the same irrigation canal and were located next to one another throughout the lower Moquegua Valley. Ilo-Tumilaca style ceramics are clearly similar to their middle valley Tumilaca counterparts (Owen 1993: 17, 190–191, 2005: 66) and over time became Ilo-Cabuza, which are simpler in design and poorer in execution than the earlier Ilo-Tumilaca ceramics. Owen (1993: 18–19) claims that following their arrival there was a progressive decrease in both the number and size of ITC sites and that ITC sites and ceramics were abandoned by AD 1250, thereby leaving the Chiribaya as the only cultural presence on the coast of the lower Moquegua Valley.

Based upon his systematic examination of ceramics excavated from both tombs and stratified deposits at the San Gerónimo, Chiribaya Alta, and Chiribaya Baja sites of the lower Moquegua Valley, David Jessup (1990a, 1990b) developed a three-phase relative chronology for Chiribaya ceramics; Chiribaya-Algarrobal phase ceramics are the earliest, followed by the

Chiribaya-Yaral phase, with the Chiribaya–San Gerónimo phase being the latest. Jessup (1990a: 27) emphatically states that tombs with the Chiribaya-Yaral type of ceramics are *always* found stratigraphically below those with Chiribaya–San Gerónimo ceramics. Others (such as García 1988 and Owen 1993) have also reported similar sequences for Chiribaya ceramics, using diagnostic sherds from stratigraphically excavated deposits. Subsequently, an additional "terminal Chiribaya" style has been defined that exhibits continuity with previous Chiribaya–San Gerónimo ceramics but postdates the collapse of the Chiribaya polity sometime around AD 1350, when an El Niño event destroyed the Chiribaya irrigation canal (Owen 1993, 2005).

Ranges for calibrated bone collagen dates reported by María Lozada (1998: 52; Lozada and Buikstra 2002: 54–57) and Jane Buikstra et al. (2005: 73), however, date the Chiribaya between AD 680 and 1250. According to these investigators, the bone collagen dates establish an early Middle Horizon emergence for the Chiribaya culture that overlapped with the Tiwanaku colonies in the middle Moquegua Valley. Therefore, by extension, the coastal Chiribaya could not have their origins among the Tiwanaku colonists in the middle Moquegua Valley. These bone collagen date ranges are centuries earlier than numerous dates reported for other Chiribaya contexts (Owen 1993: 407–408, 2002, 2005: 58) and contradict relative chronologies derived from both ceramic seriation and excavated stratified sequences (García 1988; Jessup 1990a; Owen 1993). Further, Owen (2002) has established that the marine reservoir effect is greater for the Moquegua coastal region than for many other regions of the world. His investigation of the relationship between radiocarbon dates from both Chiribaya human remains and their associated wool textiles indicates that dates based upon the human remains are slightly more than 100 years earlier than those from their associated textiles (Owen, personal communication, 2007). Given studies indicating that the Chiribaya had a large marine component in their diet (Owen 1993; Sandness 1992; Tomczak 2003), Owen (2002) suggests a dramatic carbon reservoir effect for dates from human bone collagen. Most of both the calibrated radiocarbon dates based upon nonhuman samples and excavations of stratified sequences corroborate Jessup's (1990a) ceramic seriation for the Chiribaya and chronologically place the origins of the Chiribaya-Algarrobal style as early as AD 890; Chiribaya-Yaral phase ceramics roughly date to AD 1100–1200, and the Chiribaya–San Gerónimo style was used between AD 1200 and 1390 (Owen 2005: 58). Until stratified sequences and non-bone collagen dates provide independent validation for the radiocarbon date ranges presented by Lozada (1998; Lozada and Buikstra 2002) and Buikstra et al. (2005), I will rely upon the absolute chronology first established by others.

Ethnogenesis and Biodistance Analysis in the Moquegua Valley

Materials and Methods

Dental trait data for a total of 14 mortuary samples representing nearly 900 individuals are examined in this study (table 5.1). In addition to the indigenous Ilo Preceramic and three post-Tiwanaku Chiribaya samples from the Moquegua Valley, I also analyzed an Andean Paleoindian sample, a Tiwanaku sample from the Lake Titicaca region, and eight samples representing all periods from the nearby Azapa Valley, Chile (figure 5.1). Further discussion of sample composition can be found in previous publications (Sutter 2000, 2005, 2006, in press).

For each sample I visually inspected and scored 31 dental morphological traits using the Arizona State University Dental System (Turner et al. 1991). I then calculated C. A. B. Smith's Mean Measure of Divergence (MMD) to estimate genetic relatedness among the mortuary samples examined here and tested for statistical significance (Harris and Sjøvold 2004). Although they are nonmetric distance measures, MMD distances are generally highly correlated with metric biodistance statistics and are useful for model-free

Table 5.1. Mortuary Samples Examined in This Study

Mortuary Sample	Sites	Sample Size	Period
1. Paleoindian	Throughout Andes	34	Pre-8000 BC
2. Tiwanaku	Isla del Sol, Tiwanaku	57	AD 200–750
Moquegua Valley			
3. El Yaral	La Yaral	52	AD 900–1200
4. Ilo Preceramic	Yara, Kilometer-4	16	3000–1000 BC
5. Chiribaya Alta	Chiribaya Alta	185	AD 900–1350
6. San Gerónimo	San Gerónimo	57	AD 1000–1350
Azapa Valley			
7. Chinchorro	Morro-1, 5, 6, Playa Miller-8	84	5000–2000 BC
8. Playa Miller-7	Playa Miller-7 (El Laucho)	62	500 BC–AD 500
9. Alto Ramírez	Azapa-14, 70, and 114	72	500 BC–AD 500
10. Azapa-6	Azapa-6 (Cabuza)	45	AD 900–1350
11. Azapa-71	Azapa-71 (Cabuza)	64	AD 900–1350
12. Azapa-140	Azapa-140 (Maitas-Chiribaya)	82	AD 900–1350
13. Playa Miller-4	Playa Miller-4 (San Miguel)	43	AD 1100–1450
14. Azapa-8	Azapa-8 (Gentilar)	28	AD 1300–1450

pattern recognition approaches such as that adopted here. Males and females were pooled to increase sample sizes. Twenty traits were excluded due to either sexual dimorphism or inadequate sample size.

To evaluate the biological relationships among populations in the South Central Andes two approaches are presented; these approaches both utilize the matrix of intersample biodistance statistics generated from the dental morphological variables. The first analyzes the MMD matrix using hierarchical cluster procedures that produce two-dimensional tree diagrams of nested groupings of the phenetic relations among the samples being compared. This provides a visual depiction of the relationships between samples. A more formal approach is offered by the use of design matrices. I developed two hypothetical design matrices that reflect alternative hypotheses for the origins of the Moquegua Chiribaya (*coastal señorío* and *two-stage diaspora*). The details of each design matrix are discussed below; however, each reflects intersample distances based on spatial and temporal variables. In other words, the design matrices represent both spatial and temporal lag distances between samples such that a very recent and a very ancient sample that are spatially discrete will have a large value in the design matrix. The design matrix that represents true historical relationships should be positively correlated with the matrix of biodistances. Statistical significance is evaluated by computing 999 random permutations using Peter Smouse et al.'s (1986) extension of the Mantel test. This partial Mantel test permits one to determine both the correlation and level of significance between the observed biodistance matrix and one of the design matrices, while removing the effects of the other design matrix.

Hypothetical Design Matrices

The coastal *señorío* model (figure 5.2) proposes that the Chiribaya are an indigenous coastal population that was directly descended from Early Ceramic coastal populations (Buikstra et al. 2005; Lozada 1998; Lozada and Buikstra 2002), while the two-stage diaspora model (figure 5.3) asserts that the Chiribaya represent descendants of Tiwanaku colonists who arrived on the coast as part of a population dispersal following the collapse of Tiwanaku's influence in the middle Moquegua Valley (Goldstein 2005; Owen 1993, 2005; Sutter 2000, in press). The models I employ here largely follow those that I have previously developed to examine relationships between the Tiwanaku and the prehistoric Azapa Valley samples also used here (Sutter 2006). For both the coastal *señorío* and two-stage diaspora models, I use the design matrix values that I previously described for a divergent gene flow

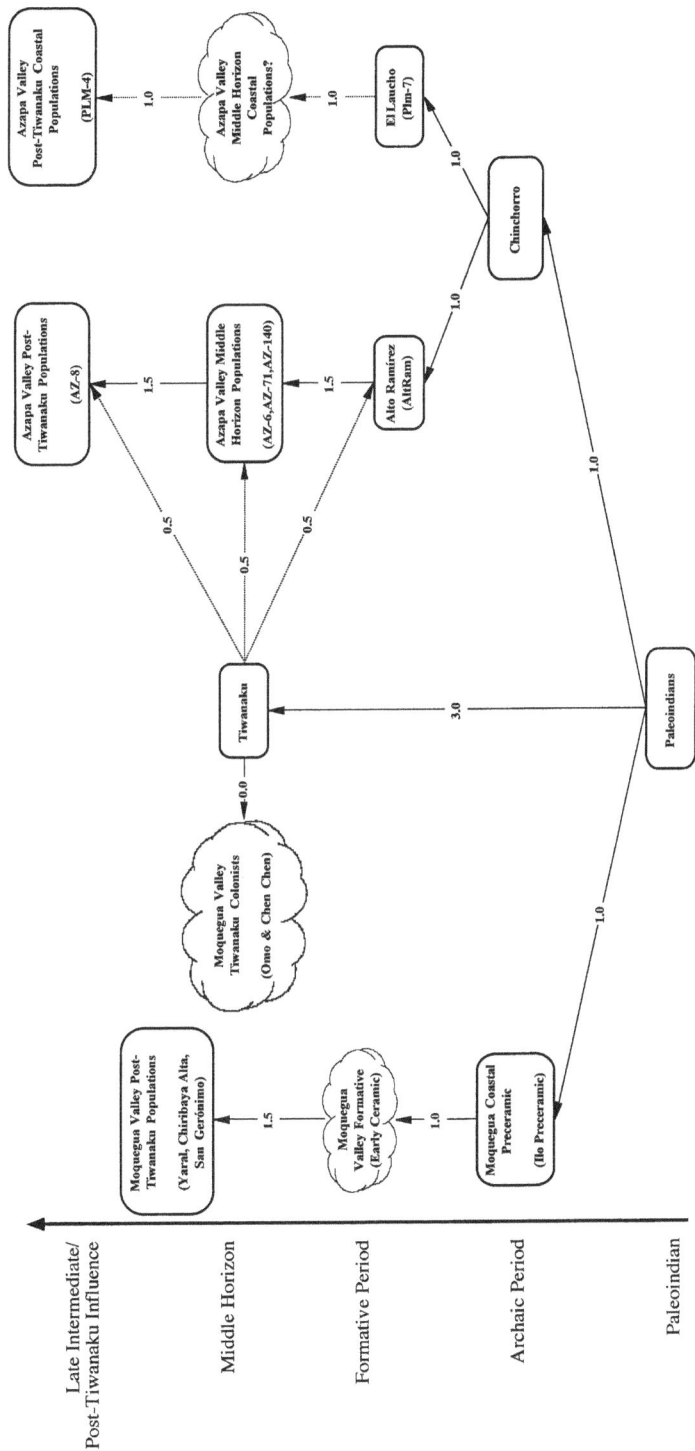

Figure 5.2. Graphic depiction of the design matrix for the coastal *señorío* model of indigenous origins for the post–Tiwanaku influence Chiribaya of the Moquegua Valley. Accordingly, Chiribaya social identity represents an entire suite of both conscious and unconscious Tiwanaku stylistic characteristics that were adopted and modified by indigenous coastal peoples.

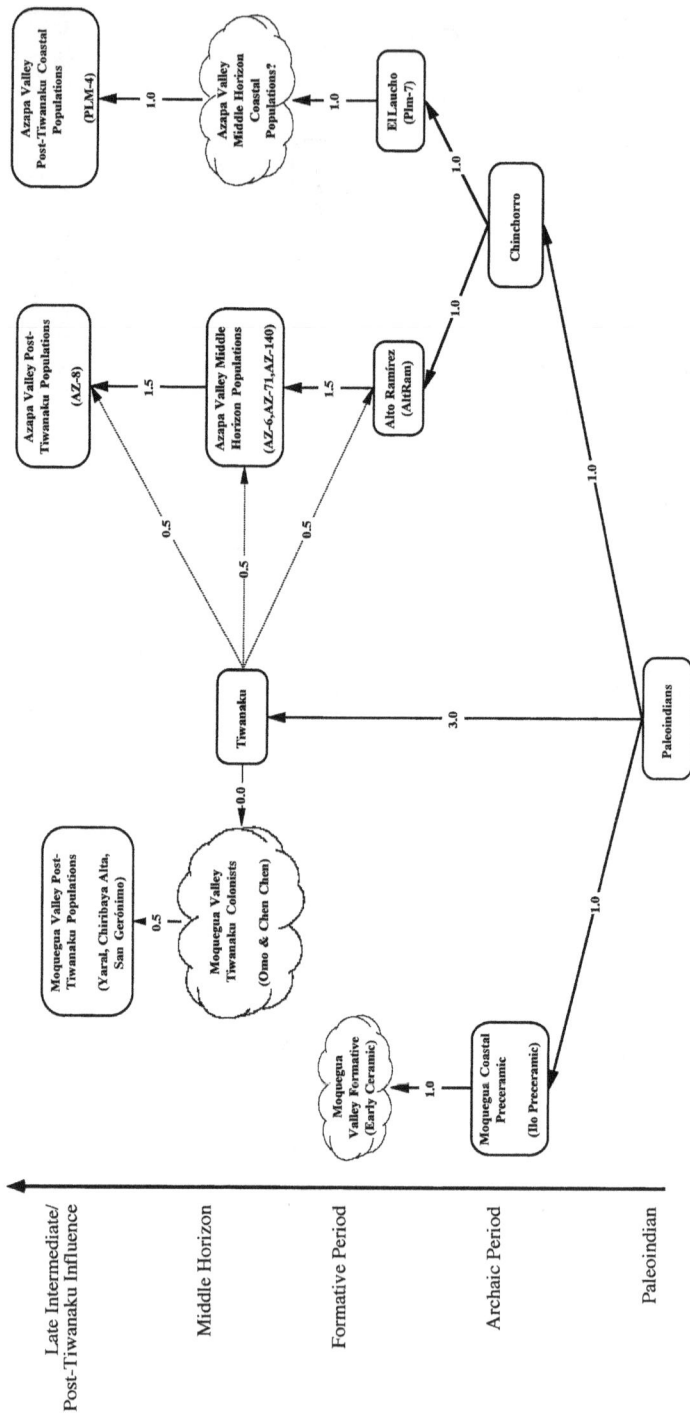

Figure 5.3. Graphic depiction of the design matrix for the two-stage diaspora model of middle valley origins for the post-Tiwanaku influence Chiribaya of the Moquegua Valley. Accordingly, the Chiribaya represent former Tiwanaku colonists who dispersed to the coast and developed a new social identity.

model for hypothetical distances among the Paleoindian, Tiwanaku, and Azapa Valley mortuary samples (Sutter 2006: figure 3). The coastal *señorío* and two-stage diaspora models also share the branching point following initial colonization of the region by the Paleoindians; both the contemporaneous and culturally similar Archaic Period Ilo Preceramic samples and Chinchorros are assigned a value of "1" for their temporal distance from the Paleoindians. Therefore, the only design matrix values that differ between the coastal *señorío* and two-stage diaspora models are those for hypothetical distance comparisons made with the four Moquegua Valley samples. For the coastal *señorío* model (figure 5.2), Moquegua samples are hypothesized to be directly descended from earlier indigenous coastal populations—such as the Ilo Preceramic sample examined in this study. Each chronological period between the Ilo Preceramic and post-Tiwanaku Chiribaya mortuary samples is assigned a value of "1" to account for temporal changes between the samples.

The two-stage diaspora model (figure 5.3) posits that the Moquegua Chiribaya represent direct descendants of former middle valley Tiwanaku colonists who dispersed to the coast, where they developed a new social identity. Based upon the assumption that the Tiwanaku received little genetic contribution from coastal peoples of the South Central Andes, they are assigned a hypothetical distance of "3" from the founding Paleoindians. Given other research indicating that the middle valley Tiwanaku colonists likely came from Tiwanaku, they are assigned a distance of "0," while descendants of the post-Tiwanaku colonists, such as the coastal Chiribaya, are assigned a value of "0.5" due to their temporal proximity to both the Tiwanaku colonists and the Tiwanaku themselves.

Additional biodistances were calculated for Moquegua Valley samples by cemetery; for Chiribaya Alta, no dental data were collected for Chiribaya Alta cemetery 8, while cemeteries 5 and 9 exhibited an unacceptably low number of observations. Thus the options were either to drop dental trait observations from these two cemeteries from subsequent analyses or to combine observations for cemeteries 5 and 9 with those from spatially proximate and stylistically similar cemeteries. Analyses were conducted both without data from Chiribaya Alta cemeteries 5 and 9 and by combining the dental trait frequencies of both cemeteries 5 and 6 and cemeteries 1 and 9. Combining the data did nothing to change how cemeteries 1 and 6 clustered with the other mortuary samples when 5 and 9 were added to the analyses. Therefore I present the resulting cluster diagram based upon biodistances calculated using cemeteries 1/9 and cemeteries 5/6 as mortuary samples.

Results of Biodistance Analysis

Inspection of the biodistance matrix (table 5.2) shows that there are relatively few significant comparisons among the 14 Andean mortuary samples examined, indicating that South Central Andean populations exhibit relatively little phenetic variation. The cluster diagram for the biodistances indicates that two groups are present (figure 5.4). The first cluster contains both the Tiwanaku and all three Chiribaya mortuary samples from the Moquegua Valley, while the second cluster consists of the Paleoindians, the Ilo Preceramic sample, and all eight of the mortuary samples from the Azapa Valley. I have previously reported similar biodistance results in a variety of

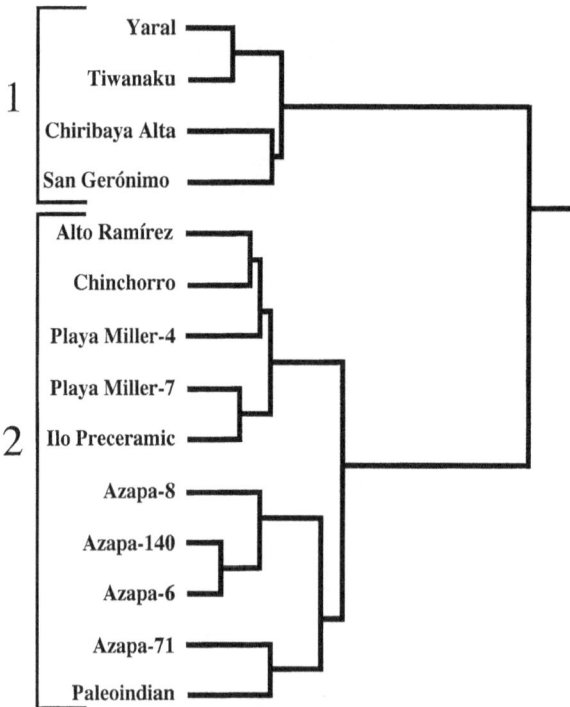

Figure 5.4. Hierarchical cluster diagram of dentally derived biodistances. Within the diagram, lower Moquegua Valley Chiribaya mortuary samples are phenetically more similar to the sample from Tiwanaku than to the Paleoindians, Azapa Valley samples, and founding Archaic Period Ilo Preceramic sample from the coastal Moquegua Valley.

Table 5.2. Biodistance Values for Prehistoric South Central Andean Mortuary Samples

	Pale	Tiw	Yar	IIPC	ChA	SG	Chin	Plm7	Plm4	AlRm	A140	A6	A71	A8
Paleoindian	0.00													
Tiwanaku	0.76	0.00												
El Yaral	1.37	-0.39	0.00											
Ilo Preceramic	0.39	1.40	1.43	0.00										
Chiribaya Alta	0.55	0.11	0.43	0.75	0.00									
San Gerónimo	1.41	0.35	0.16	1.12	0.31	0.00								
Chinchorro	0.36	2.59	4.22	0.01	2.81	2.69	0.00							
Playa Miller-7	0.82	3.39	4.53	-0.31	3.37	2.89	0.10	0.00						
Playa Miller-4	0.75	1.75	2.70	0.30	1.69	1.28	-0.08	0.26	0.00					
Alto Ramirez	0.31	1.79	2.97	0.32	1.44	1.29	-0.12	0.33	0.11	0.00				
Azapa-140	0.82	1.20	1.72	0.17	1.03	0.70	1.53	1.39	1.29	0.51	0.00			
Azapa-6	0.51	0.82	1.08	0.08	0.35	0.46	1.10	1.19	1.01	0.54	-0.54	0.00		
Azapa-71	0.20	1.80	2.36	-0.12	1.12	1.51	1.00	1.03	1.13	0.62	0.51	0.05	0.00	
Azapa-8	1.05	0.39	0.54	0.29	0.26	0.21	1.19	1.36	0.93	0.50	-0.11	0.00	0.48	0.00

publications (Sutter 2000, 2005, in press). While MMD values provide an indication as to the relative biological distance and associated significance among the mortuary samples being compared, however, their interpretation is often equivocal and may not indicate which among a number of competing explanations is best supported by the observed biodistances. Therefore the observed MMD values were compared to the two prevalent models for Chiribaya origins, using the hypothetical matrix model.

Comparisons of the observed biodistances with the two hypothetical model matrices using Smouse et al.'s (1986) partial Mantel test provide support for the results obtained by hierarchical clustering. The two-stage diaspora model produces a statistically significant correlation with observed biodistances ($r = 0.320$, $p = 0.008$), while the coastal *señorío* model produces a small and nonsignificant correlation ($r = 0.080$, $p = 0.215$) with the observed biodistance matrix.

Analysis of each spatially distinct cemetery at Chiribaya Alta and La Yaral with the Paleoindian, Tiwanaku, Ilo Preceramic, and San Gerónimo samples also reveals interesting patterns. Inspection of the hierarchical cluster diagram of cemetery samples for the Moquegua Valley Chiribaya produces two clusters (figure 5.5). The first cluster contains the Tiwanaku and San Gerónimo samples, both samples from La Yaral, and Chiribaya Alta samples from cemeteries 2, 3, 5/6, and 7, while the second contains both the Paleoindian and Ilo Preceramic samples and Chiribaya Alta mortuary samples from cemeteries 1/9 and 4.

Chiribaya Ethnogenesis: Discussion and Conclusions

The results of both biodistance and matrix correlation analyses from this study support the two-stage diaspora model for explaining the genetic and cultural origins of the coastal Chiribaya of the Moquegua Valley, whereas there was a poor fit and a low and nonsignificant correlation with the coastal *señorío* model. Importantly, these results would be unaffected if Lozada's (1998) and Buikstra et al.'s (2005) chronology for the Chiribaya is correct. The treatment of Moquegua Valley cemeteries as discrete mortuary samples, however, provides a more subtle interpretation of how the Chiribaya may have emerged and persisted as a distinct and perhaps multiethnic polity in the wake of the political vacuum associated with the collapse of both the Tiwanaku and Wari influences; while formerly Tiwanaku Omo- and Chen Chen–style peoples were primarily responsible for Chiribaya ethnogenesis,

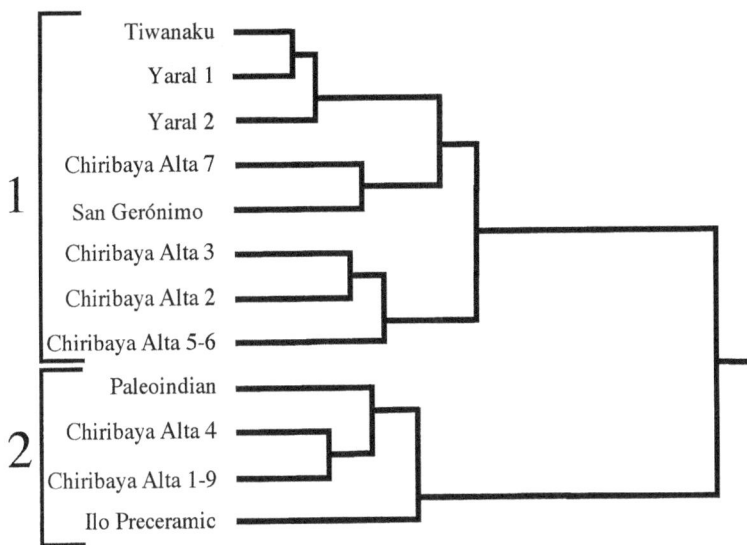

Figure 5.5. Hierarchical cluster diagram of dentally derived biodistances for comparisons of Moquegua Valley Chiribaya cemetery samples with Tiwanaku, Ilo Preceramic, and Paleoindian mortuary samples. While Chiribaya mortuary samples from La Yaral and Chiribaya Alta cemeteries 2, 3, 5–6, and 7 are phenetically more similar to the Tiwanaku sample, those from Chiribaya Alta cemeteries 1–9 and 4 are more similar to the Paleoindian and Ilo Preceramic samples. These results imply that Chiribaya ethnic identity was forged by both formerly Tiwanaku colonists from the middle Moquegua Valley and indigenous coastal peoples.

it appears that indigenous coastal peoples descended from the Early Ceramic tradition may also have contributed to Chiribaya ethnic identity.

Both the archaeological and biodistance data considered here clearly point to the Chiribaya having their cultural and genetic origins among earlier Tiwanaku colonists of the middle Moquegua Valley, rather than representing an exclusively coastal cultural phenomenon with earlier coastal antecedents. Contrary to Lozada's (1998: 64) assertion that only ceramics have been used to infer genetic relations between the Tiwanaku and Chiribaya colonists of the middle valley, Owen (1993, 2005: 66) has cogently argued that nearly the entire suite of cultural characteristics associated with both the Chiribaya and Ilo-Tumilaca/Cabuza traditions has no known antecedents in the coastal region of the Moquegua Valley. Lozada (1998: 112, 177; Lozada and Buikstra 2002: 102–103) claims that both the Chiribaya and Ilo-Tumilaca represent indigenous coastal peoples who simply adopted middle

valley Tiwanaku cultural characteristics. I would agree that genetically unrelated people sometimes do adopt some aspects of the emblemic, assertive stylistic elements from a different ethnic group, especially if they are
actively attempting to realign their ethnic identity. But I find it problematic
to suggest that genetically unrelated peoples would adopt *both* the assertive and unconscious (i.e., isochrestic variation) aspects of material culture
(Sackett 1982) of another ethnic group without substantial gene flow such
as intermarriage between them. Both assertive and isochrestic elements of
ethnicity emerge from *habitus* (Bourdieu 1977), and, more specifically, the
acquisition of knowledge regarding post-Tiwanaku unconscious, isochrestic cultural variation would primarily have occurred through practice that
comes through intimate contact among peoples in the form of enculturation and intermarriage (Shennan 1989: 19).

Owen (2005: 58) reports that the latest dates for the middle valley Chen
Chen style are around AD 1000 and coincide with the appearance and dispersal of both upper valley Tumilaca and lower valley Ilo-Tumilaca traditions. The radiocarbon dates reported by Owen (2002, 2005) provide evidence for approximately one century of overlap following the arrival on the
coast of users of Chiribaya-Algarrobal ceramics while Omo-style and Chen
Chen–style settlements in the middle Moquegua Valley were still occupied.
The collapse of the ethnically Tiwanaku settlements in the middle valley
probably did not occur overnight, however, so some overlap is to be expected if the Chiribaya-Algarrobal arrived during the dispersal of Tiwanaku
colonists. Indeed, Williams (2002) suggests that lower-elevation Omo-style
communities of the middle Moquegua Valley likely dispersed first, shortly
after Chen Chen–style settlements upstream scrambled to modify their irrigation canals in the wake of hydrological works by the Wari colonies located in the upper reaches of the middle valley. This suggests a gradual and
progressive abandonment of the Omo-style then Chen Chen–style middle
valley settlements.

While ceramics of the earliest Chiribaya-Algarrobal phase share pastes,
slipping techniques, temper, colors, and some of their forms and design elements with the Ilo-Tumilaca (Owen 1993: 190–191, 2005: 66), the cultural
antecedents of Chiribaya-Algarrobal ceramics remain enigmatic. Despite
these similarities, there are no clear middle valley antecedents for Chiribaya-
Algarrobal ceramics (Owen 1993: 191). Did the users of Chiribaya-Algarrobal ceramics represent formerly Omo-style Tiwanaku colonists who were
the first to disperse? Given the clear continuity between Chen Chen–style
and subsequent Tumilaca and Ilo-Tumilaca ceramics, could it be that the

migratory herding Omo-style peoples became dependent upon the more stationary Chen Chen–style settlements to provide them with their ceramics as the Omo-style people tried to find a new location to live and develop a new ethnic identity? Chiribaya cultural origins among people who used Omo styles might also explain the importance of textiles and camelids to the Chiribaya, as evidenced by the preponderance of camelid remains scattered across Chiribaya Alta and the inclusion of camelid parts (including hooves, skulls, and complete animals) in Chiribaya graves (Owen 2005); this practice has not been reported among any Middle Horizon population from the middle (Goldstein 2005: 251) or lower valley (Owen 1993).

I am not yet willing to argue for such a provocative origin for the Chiribaya, but this model has appeal because it would presumably be testable with the future study of available mortuary samples from the middle valley and possible sourcing analyses for ceramics in the Moquegua Valley. Furthermore, such a model is not completely without merit; Owen (1993: 192, 2005: 56) mentions the possibility that ceramics of both the Ilo-Tumilaca and Chiribaya-Algarrobal phases were manufactured by the same ceramicists. Goldstein (2005: 217) reports that marine foods compose a small yet important portion of the Middle Horizon diets at Omo-style and Chen Chen–style sites. If his characterization of Omo-style settlements consisting of migratory camelid-herding merchants is correct, then it is likely that Omo-style merchants were the importers of marine resources to the middle Moquegua Valley during Tiwanaku's influence there. This may have set up a situation where the Omo-style merchants established Early Ceramic tradition clients on the coast, thereby facilitating their move to the lower Moquegua Valley. Owen's (1993, 2005) survey data and excavations in the lower Moquegua Valley indicate that the region had relatively few sparsely populated settlements occupied by Early Ceramic tradition peoples. The sparsely populated coastal region where the formerly Omo-style herders already had established trade partners would have made an inviting location for settlement (Owen 1993: 125).

Conversely, did Omo-style peoples' ties to the Copacabana Peninsula become strained? If, as Goldstein (2005: 151) has suggested, Omo-style ceramics were imported from the Tiwanaku heartland, then the users of Omo-style ceramics may have needed a new source of feasting vessels after losing access to their *altiplano* source for Omo-style ceramics. Goldstein (2005: 312) fittingly states that "most long-distance migrations are motivated by a combination of 'push' and 'pull' factors that create a perception of difference in economic productivity between homelands and frontiers."

While this scenario may be speculative, it could explain why formerly Omo-style people could make such an abrupt switch to a new ceramic style (i.e., Chiribaya-Algarrobal ceramics) with strong technical similarities yet stylistically distinct designs in comparison to those of the Tumilaca and Ilo-Tumilaca traditions, which presumably would have been made by descendants of the Chen Chen–style peoples. In other words, formerly Omo-style peoples were not initially the manufacturers of Chiribaya-Algarrobal ceramics—the Tumilaca (i.e., formerly Chen Chen–style peoples) were. Further, once the peoples of the Chiribaya-Algarrobal phase became established in the lower valley (whether from an Omo-style or some as yet unidentified formerly Tiwanaku ethnic group), it should come as no surprise that they would have continued to maintain ties with other ethnically Tiwanaku and post-Tiwanaku middle valley peoples with whom they once peacefully shared crucial resources and political affiliations. Such relations could have facilitated the subsequent arrival (Anthony 1990) and peaceful coexistence of Ilo-Tumilaca peoples some five to ten decades later (Owen 2005: 59, 72). This relationship would make sense if these Tumilaca and Chiribaya-Algarrobal peoples shared exchange relations for complementary resources, such as initially trading ceramics and middle valley resources in exchange for marine and other lower valley resources.

Biodistance analyses for Moquegua Valley Chiribaya cemetery samples suggest, however, that Chiribaya ethnogenesis was a complex process involving both post-Tiwanaku migrants and indigenous coastal peoples. In particular, the cemetery data from Chiribaya Alta accord well with both DNA (Haydon 1993) and stable isotope analyses (Knudson and Price 2007) indicating that those interred from Chiribaya Alta were genetically variable and possibly raised in a variety of locations. The Chiribaya Alta cemetery 1/9 sample represents individuals interred with early Chiribaya-Algarrobal ceramics and is found in the same cluster as earlier Ilo Preceramic and Paleoindian mortuary samples, while the cemetery 4 sample is also found in the same cluster and primarily represents individuals associated with ceramics of the Chiribaya–San Gerónimo phase. The remaining Chiribaya samples (San Gerónimo, La Yaral cemeteries 1 and 2, and Chiribaya Alta cemeteries 2, 3, 5/6, and 7) all cluster with the Tiwanaku sample. These results indicate that some Chiribaya peoples almost certainly did have some indigenous coastal ancestry that may have been maintained throughout the existence of the Chiribaya polity. This interpretation is not without problems, as it does not closely correspond with the grave goods found in Chiribaya Alta cemeteries 1 and 9, with implements indicating agropastoral subsistence, while grave furniture for Chiribaya Alta cemetery 4 is consistent with marine subsistence activities (Lozada 1998; Lozada and Buikstra 2002). Intermarriage of indigenous coastal peoples with immigrant middle valley agropastoral peoples may have blurred the strict inheritance implied by Lozada and Buikstra for ethnically associated occupational roles, however, especially since many occupational tasks would have been gender specific.

Owen (1993, 2005) has suggested that the arrival of peoples associated with the ITC tradition peoples probably began after the establishment of the lower valley Chiribaya population. This may account in part for their relatively impoverished graves and inferior ceramics. Owen (2005: 66–70) documents significant differences in the occupationally related grave furniture of ITC and Chiribaya tombs; while Chiribaya tombs often contain model boats, fishing tackle, and botanical offerings from the nearby fog oases, these items are absent from the graves of ITC burials, which tend to reflect agropastoral activities. Notably, there is also a dramatic difference in both the quality and number of offerings, with the Chiribaya tombs often having two to five times the number of grave offerings interred in ITC tombs. After the initial arrival of peoples using Ilo-Tumilaca ceramics, the number, size, and intensity of Ilo-Cabuza occupations decreased, with the ITC tradition finally "disappearing" sometime around AD 1250. Simultaneously, the size, number, and intensity of Chiribaya settlements increased. Rather than indicating a dispersal of ITC peoples, the increasingly impoverished and dwindling number of ITC peoples may have played an integral role in the making of post–Chiribaya-Algarrobal ethnic identity through intermarriage.

What, then, can be said of Chiribaya ethnogenesis? In this chapter I have reported biodistance analyses that support the two-stage diaspora archaeological model suggesting that the Chiribaya were largely descended from middle valley Tiwanaku colonists. While some individuals interred at Chiribaya Alta likely had indigenous coastal roots, archaeological evidence indicates that the ITC peoples also contributed to the Chiribaya's genetic composition. But the Chiribaya's genetic origins do not provide us with a complete understanding of how their ethnic identity emerged.

Some have noted that ethnic shifts can occur when individuals feel "estranged and isolated from a distracted postindependence" (Haley and Wilcoxon 2005: 433) polity or when "identity lost salience amidst changing conditions, as subjects sought higher status, or because of a combination of these" (Haley and Wilcoxon 2005: 441). I would argue that both explanations played a role in the post-Tiwanaku Moquegua Valley. The material expressions of new identities are often constrained by history and tradition (Bawden 2005) yet reflect a rejection of previous assertive style associated with the former symbols of power and authority: the monumental architecture and Gateway God that were once powerful symbols of the Tiwanaku drop out of post-Tiwanaku peoples' repertoire of material expressions of group identity. While many unconscious expressions of ethnicity (such as mortuary practices, grave offerings, textile manufacture, and ceramic forms) continued with only minor modifications, a new stylistic set of abstract geometric designs appeared and then persisted on Chiribaya ceramics and textiles for nearly four centuries.

Early during the emergence of Chiribaya ethnic identity, others with whom they had economically and socially complementary relations also contributed to Chiribaya group identity. Accordingly, indigenous coastal peoples contributed to the new ethnic groups' marine resources, and users of Chiribaya-Algarrobal ceramics (formerly Omo-style people?) contributed textiles and camelids to exchange with still-friendly post-Tiwanaku clients, while Ilo-Tumilaca peoples initially contributed agricultural resources to the economy.

Finally, Barth (1969) explains that individuals can and do cross ethnic boundaries when it behooves them. What factors influenced others, such as the Ilo-Tumilaca/Cabuza, to surrender their own ethnic identity and become Chiribaya? I suggest that the answer is found in the asymmetric nature of their economic relations with the Chiribaya, as evidenced by both the quality and quantity in grave offerings and domestic structures (Owen 2005: 70, 72–73). This disparity arose due to the ITC peoples' late entrance to the lower valley; the Chiribaya were probably already established and beginning to prosper by the time the ITC peoples arrived. Indeed, knowledge of Chiribaya prosperity may have been the initial attraction for the ITC peoples if they perceived greater economic opportunities available in the lower valley. As Chiribaya economic success increased relative to ITC peoples, however, this disparity provided a powerful material motivation for the ITC peoples to give up an economically less-successful ethnic identity. Indeed, while Barth (1969) suggests that ethnic identity emerges through a process of negotiating both the local and external social and environmental landscapes—a process of becoming an ethnic individual that Bourdieu (1977) refers to as *habitus*—Gunnar Haaland (1969: 71) observes that individual ethnic shifts result from practicing the subsistence activities of other ethnic groups. Accordingly, over time, the Moquegua Valley Chiribaya, who began as a polity of economically defined ethnic groups, attracted others, such as the Ilo-Tumilaca/Cabuza people, to develop a single ethnic identity through economic prosperity and intermarriage.

References Cited

Anthony, David W. 1990. "Migration in Archaeology: The Baby and the Bathwater." *American Antiquity* 92: 895–914.

Barth, Fredrik. 1969. "Introduction." In *Ethnic Groups and Boundaries: The Social Organization of Culture Difference*, edited by F. Barth, 9–38. Boston: Little, Brown and Company.

Bawden, Garth. 2005. "Ethnogenesis at Galindo, Peru." In *Us and Them: Archaeology and Ethnicity in the Andes*, edited by R. M. Reycraft, 12–33. Los Angeles: Cotsen Institute of Archaeology, University of California, Los Angeles.

Blom, Deborah E., Benedikt Hallgrímsson, Linda Keng, María C. Lozada C., and Jane E. Buikstra. 1998. "Tiwanaku 'Colonization': Bioarchaeological Implications for Migration in the Moquegua Valley, Peru." *World Archaeology* 30: 238–261.

Bourdieu, Pierre. 1977. *Outline of a Theory of Practice*. Cambridge: Cambridge University Press.

Buikstra, Jane E., Paula D. Tomczak, María C. Lozada Cerna, and Gordon F. M. Rakita. 2005. "Chiribaya Political Economy: A Bioarchaeological Perspective." In *Interacting with the Dead: Perspectives on Mortuary Archaeology for the New Millennium*, edited by G. F. M. Rakita, J. E. Buikstra, L. A. Beck, and S. R. Williams, 66–80. Gainesville: University Press of Florida.

García, Manuel. 1988. "Excavaciones de dos viviendas Chiribaya en el Yaral, valle de Moquegua." Thesis, Universidad Católica "Santa María," Arequipa, Peru.

Goldstein, Paul S. 2005. *Andean Diaspora: The Tiwanaku Colonies and the Origins of South American Empire*. Gainesville: University Press of Florida.

Haaland, Gunnar. 1969. "Economic Determinants in Ethnic Processes." In *Ethnic Groups and Boundaries: The Social Organization of Culture Difference*, edited by F. Barth, 58–73. Boston: Little, Brown and Company.

Haley, Brian D., and Larry R. Wilcoxon. 2005. "How Spaniards Became Chumash and Other Tales of Ethnogenesis." *American Anthropologist* 107: 432–445.

Harris, Edward F., and Torstein Sjøvold. 2004. "Calculation of Smith's Mean Measure of Divergence for Intergroup Comparisons Using Nonmetric Data." *Dental Anthropology* 17: 83–93.

Haydon, Rex C. 1993. "Survey of Genetic Variation among the Chiribaya of the Osmore Drainage Basin, Southern Perú." Doctoral dissertation, University of Chicago, Department of Anthropology.

Jessup, David. 1990a. "Desarrollos generales en el intermedio tardío en el valle de Ilo, Perú" (Internal Report for Programa Contisuyu). Submitted to Informe Interno de Programa Contisuyu.

———. 1990b. "Rescate arqueológico en el museo de sitio de San Gerónimo, Ilo." In

Trabajos arqueológicos en Moquegua Perú: Volumen 3, edited by L. K. Watanabe, M. E. Moseley, and F. Cabieses, 151–165. Moquegua: Programa Contisuyu del Museo Peruano de Ciencias de la Salud/Southern Peru Copper Corporation.

Knudson, Kelly J., and T. Douglas Price. 2007. "Utility of Multiple Chemical Techniques in Archaeological Residential Mobility Studies: Case Studies from Tiwanaku- and Chiribaya-Affiliated Sites in the Andes." *American Journal of Physical Anthropology* 132: 25–39.

Lewis, Cecil M., Jr., Jane E. Buikstra, and Anne C. Stone. 2007. "Ancient DNA and Genetic Continuity in the South Central Andes." *Latin American Antiquity* 18: 145–160.

Lozada, María C. 1998. "The Señorío of Chiribaya: A Bio-Archaeological Study in the Osmore Drainage of Southern Peru." Doctoral dissertation, University of Chicago, Department of Anthropology.

Lozada, María C., and Jane E. Buikstra. 2002. *El Señorío de Chiribaya en la Costa Sur del Perú*. Lima: Instituto de Estudios Peruanos.

Owen, Bruce. 1993. "A Model of Multiethnicity: State Collapse, Competition, and Social Complexity from Tiwanaku to Chiribaya in the Osmore Valley, Perú." Doctoral dissertation, University of California, Los Angeles, Department of Anthropology.

———. 2002. "Marine Carbon Reservoir Age Estimates for the Far South Coast of Peru." *Radiocarbon* 44: 701–708.

———. 2005. "Distant Colonies and Explosive Collapse: The Two Stages of the Tiwanaku Diaspora in the Osmore Drainage." *Latin American Antiquity* 16: 45–80.

Sackett, James R. 1982. "Approaches to Style in Lithic Archaeology." *Journal of Anthropological Archaeology* 1: 59–112.

Sandness, Karin. 1992. "Temporal and Spatial Dietary Variability in the Osmore Drainage, Southern Peru: The Isotope Evidence." Master's thesis, University of Nebraska at Lincoln, Department of Anthropology.

Shennan, Stephen. 1989. "Introduction: Archaeological Approaches to Cultural Identity." In *Archaeological Approaches to Cultural Identity*, edited by S. Shennan, 1–32. London: Unwin Hyman.

Smouse, Peter E., Jeffrey C. Long, and Robert R. Sokal. 1986. "Multiple Regression and Correlation Extensions of the Mantel Test Matrix Correspondence." *Systematic Zoology* 35: 627–632.

Sutter, Richard C. 2000. "Prehistoric Genetic and Culture Change: A Bioarchaeological Search for Pre-Inka Altiplano Colonies in the Coastal Valleys of Moquegua, Peru, and Azapa, Chile." *Latin American Antiquity* 11: 43–70.

———. 2005. "The Prehistoric Peopling of South America as Inferred from Epigenetic Dental Traits." *Andean Past* 7: 183–217.

———. 2006. "The Test of Competing Models for the Prehistoric Peopling of the Azapa Valley, Northern Chile, Using Matrix Correlations." *Chungará* 38: 63–82.

———. In press. "Prehistoric Population Dynamics in the Peruvian Andes." In *The Foundations of South Highland Andean Civilization: Papers in Honor of Michael E. Moseley*, edited by P. R. Williams. Los Angeles: Cotsen Institute of Archaeology, University of California, Los Angeles.

Tomczak, Paula D. 2003. "Prehistoric Diet and Socioeconomic Relationships within the Osmore Valley of Southern Peru." *Journal of Anthropological Archaeology* 22: 262–278.

Turner, Christy G., Christopher R. Nichol, and G. Richard Scott. 1991. "Scoring Procedures for Key Morphological Traits of the Permanent Dentition: The Arizona State University Dental Anthropology System." In *Advances in Dental Anthropology*, edited by M. A. Kelly and C. S. Larsen, 13–32. New York: Wiley-Liss.

Williams, Patrick R. 2002. "Rethinking Disaster-Induced Collapse in the Demise of the Andean Highland States: Wari and Tiwanaku." *World Archaeology* 33: 361–374.

6

Surviving Contact

Biological Transformation, Burial, and Ethnogenesis in
the Colonial Lambayeque Valley, North Coast of Peru

HAAGEN D. KLAUS AND MANUEL E. TAM CHANG

The extraordinary biocultural interchange between Native Americans and Europeans beginning in the fifteenth century embodied an adaptive transition that irrevocably shaped modern humanity. Bioarchaeological studies have begun to characterize the consequences of this transition for Native Americans, revealing diverse and complex outcomes involving increased morbidity, dietary changes, and catastrophic demographic collapse, all too often interpreted as a series of passive extinctions (e.g., Crosby 1972; Haldane 2002: 37). We eschew this eschatological perspective and instead conceptualize collapse as a cyclical process of reorganization—a dynamic and creative process of sociopolitical decentralization, reorganization, and reconfiguration accompanied by ethnogenesis and the emergence of new peoples and identities (e.g., Bawden 2005; Schwartz and Nichols 2006; Stojanowski, this volume). Over the last two decades, anthropological studies of human skeletal remains have helped demonstrate that demographic collapse initiated profoundly complex biological and historical processes (Baker and Kealhofer 1996; Larsen 2001; Larsen and Milner 1994; Verano and Ubelaker 1992). The remarkable diversity in the mode, magnitude, and tempo of indigenous contact outcomes appears shaped by a dynamic interplay between European economic ambitions and precontact indigenous health status, social complexity, population structures, and ecology.

Corresponding empirical understandings of indigenous cultural changes and adaptations, however, remained largely understudied or derived from incomplete ethnohistories biased by European ethnocentrism. This lacuna leads to a number of unanswered questions about indigenous colonial experiences. How did people attempt to achieve social buffering against biological disruption? When did cultural conversion or ideological shifts occur, and how deeply did such changes penetrate? How did emic constructions of identity develop under such conditions? As ethnogenesis and identity transformations are strongly manifested within contexts of social disruption (see the discussion in Bawden 2005), bioarchaeological study of postcontact Native American identity and culture is long overdue, reflecting the relative lack of synthesis of bioarchaeology with the highly complementary social information derived from archaeology. This view is articulated by Lynne Goldstein (2006) and Rebecca Gowland and Christopher Knüsel (2006), who perceive a lack of synthetic treatment of human remains and their burial contexts. This is unfortunate. Burial practice is perhaps one of the most sensitive cultural domains manipulated by communal groups as a vehicle for ideological discourse and is conditioned directly by the social and political environment (Bawden 2005: 17). Burial patterns feature a remarkable range of encoded social information ideal for examining ethnic identity. Symbolic constructions of burial can serve as simultaneous depictions of real and desired social situations. To decode such messages, a thorough, particularistic, and deeply contextualized understanding of local history, social structure, and ritual must be established (Bawden 2001; Shimada et al. 2004)—without such context the archaeology of identity is nothing.

In this chapter we integrate funerary data with skeletal biology to provide a more holistic vision of social, historical, and biological issues. Our aim is to characterize the effects of Spanish colonialism on indigenous identity following the collapse of Andean sociopolitical systems in the ethnically Mochica town of Mórrope in the Lambayeque Valley Complex to understand how the Mochica negotiated identity throughout the colonial period. We synthesize multiple lines of complementary ethnohistoric, skeletal biological, and archaeological evidence to contextualize the colonial north coast of Peru. Profoundly asymmetrical relations of social power subjugated and exploited the Mochica into an almost stereotypical postcontact disaster as they endured socially destabilizing depopulation, decreased birthrates, markedly increased suffering from chronic diseases, and less nutritious diets. The Mochica fit the notion of a highly marginalized postcontact indigenous population at or beyond the brink of ethnocide, as their millennia-old identity would appear progressively expunged. Instead, we find biocultural evidence of ethnogenesis that occurred in two stages. First, a relatively rapid and substantial microevolutionary change appears to be the result of the expansion and redefinition of Mochica perceptions of identity and sociopolitical group boundaries. This biological transformation was followed by a series of cultural responses encoded in mortuary rituals that integrated Catholic and pre-Hispanic practices involving ritual interactions between the living and the dead related to precontact soul-transfer and fertility rituals. Burials at Mórrope reveal pre-Hispanic identities engaged in a dialectical struggle with the colonial reality, resulting in the creation of a syncretic Euro-Andean culture as the Mochica actively negotiated biosocial stress, poverty, and identity, enabling them to survive and emerge from contact as a transformed people.

Ethnicity and Identity

In this chapter we advocate a multifaceted view of identity and ethnicity, the subject of a long and rich tradition of inquiry and debate in anthropology (Barth 1969; Bentley 1987; Cohen 1974; Jones 1997). As a complex social phenomenon manifested according to circumstance and history, ethnicity perhaps can have no single definition. Our operational conception of ethnic identity draws from three complementary visions. Garth Bawden (2005: 13) states that baseline ethnicity is born from recognition of common membership in a social group that is able to reproduce itself. Richard Reycraft (2005) feels that ethnicity is the very mechanism by which individuals who share a common social identity use culture—an apparatus of social expression and change—to symbolize and enact solidarity. Siân Jones (1997: 120), who extends Pierre Bourdieu's (1977) key concept of *habitus* to the archaeology of identity, states that "[t]he construction of ethnicity, and the objectification of cultural differences that this entails, is a product of the intersection between people's habitual dispositions with concrete social conditions characterizing any given social situation." In other words, ethnic identity can be perceived as a dynamic, context-bound phenomenon in which symbols, objects, and people are enacted and manipulated on unconscious and conscious levels in dialectical relationship involving *habitus*, agency, and local historical trajectories. Ethnicity is perhaps the most dynamic of a group's social "tools," functioning as a mode of social cohesion and reproduction, a weapon of group conflict, and a foundation of group tradition, consciousness, and perception. As we demonstrate here, ethnicity is as enduring as it is ephemeral.

Regional Biocultural Context

Late Pre-Hispanic Ethnic Identity, Death, and Society

The Lambayeque Valley Complex was a center of independent cultural development on the north coast of Peru, containing the hydrologically linked

Motupe, La Leche, Lambayeque, Reque, and Zaña river valleys, spanning the Andean piedmont to the Pacific coast (figure 6.1). The Moche culture emerged on the north coast around AD 100, characterized by state-level organization (certainly by its latter phase), monumental architecture, elaborate burials, and exquisite narrative art (Bawden 1996; Pillsbury 2001; Shimada 1994). Following the Moche political collapse around AD 750, the Middle Sicán state, or Lambayeque Culture, emerged as an economically dominant, multiethnic theocracy representing the region's historical and cultural apogee from AD 900 to 1100 and produced metal objects on a scale and quality unequaled in the New World (Shimada 2000). The Late Sicán culture was conquered by the predatory Chimú state circa AD 1375, and less than 100 years later the Lambayeque region fell under the dominion of the Inka.

Burial became a potent social and ritual domain on the north coast beginning with the formative Cupisnique culture (1500–650 BC). For at least three millennia before contact, the living encoded a range of social statements about the dead and their society, including symbolisms that sanctioned the social order, maintained communal traditions, and reinforced group identity (Bawden 2001; Elera Arévalo 1998, Millaire 2002). In Middle Sicán society, institutionalized social inequality was reflected in systematic differences in burial location, tomb construction, and body disposal; differential access to gold, *tumbaga* alloy, and copper alloy grave goods reflected rank, as ethnic group boundaries were reinforced in death (Shimada et al. 2004; see also Klaus 2003).

Izumi Shimada et al. (2004) examined multiple lines of biological and artifactual evidence indicating that the small Sicán elite class was a foreign, intrusive ethnic group, possibly originating from southern Ecuador. The majority of the Middle Sicán population was composed of ethnically Mochica individuals—descendants of the earlier Moche culture (Klaus 2008). Despite changes in material culture or political organization, Mochica identity seems to have thrived among the local population just under the surface of all subsequent cultural developments in the Lambayeque region from AD 750 to 1532. The highly conservative Mochica of Lambayeque reproduced craft production technologies and distinctly Mochicoid art styles, including representations of kneeling warriors, war bundle motifs, sexualized ceramic themes, fanged Decapitator entities, and a variety of secondary icons, often hybridized with Sicán, Chimú, and Inka canons. The most significant ethnic markers are found in the conservative Mochica ritual grammar of burial that crystallized in the first few centuries AD and persisted until contact: burial

Sechura
Desert

MOTUPE RIVER

North to Piura

LA LECHE RIVER

Jayanca

Pacora

Illimo

Sicán
Precinct

Túcume

El Purgatorio

Mochumi

La Zaranda

Huaca
Letrada

HPBG Batán Grande

Cerro de
Trapos

Cerro -
Huaringa

Cerro
Blanco

La
Calera

Mórrope

Pampas de Toro
Muerto

Pitipo

Cerro
Tambo Real

Barranco
Colorado

Cerro
Plácido

Ferreñafe
Viejo

Pampa
de Chaparri

Raca Ru

Cerro
Chaname

Huaca
Sialupe

Pampas de
Mórrope

Ferreñafe

TAYMI CANAL

Pampa
de
Burros

Chotuna-
Chornancap

Lambayeque

Picsi Vista
Florida

Cerro
Patapo

Lambayeque
Viejo

Cinto

Caleta de San José

Chiclayo

LAMBAYEQUE RIVER

Santa
Rosa

Col

Colluz

REQUE RIVER

Sipán

Saltur

Cerro
Saltur

Rinconada

Cerro
Colligue

Pimentel

Reque

Cerro
de Reque

Cerro
Negro

Cerro
La Cantaria

Pampa
de Collique

Santa
Rosa

Monsefu

Cascajales

Cerro
Cerrillos

Pampa
de Reque

PAN-AMERICAN HIGHWAY

Cerro
San Nicolás

Zaña

Puerto Eten

Morro de Eten

Pampa
de Eten

Cerro
Guitarra

Playa de Lobos

Inka North-Sou

N

Pacific Ocean

0   5   10
km

Úcupe

Huaca
El Pueblo Mocupe

Pampas
de Las Delicias

The Lambayeque Valley Complex:
Major Archaeological Features

■ MAJOR MOUND,
PYRAMID, OR SITE

⊞ HABITATIONAL SITE

〜 PERENNIAL RIVER

····· SEASONAL RIVER

---▶ PREHISTORIC
CANAL

PREHISTORIC WALL

⊙ INFERRED
LATE PREHISPANIC
POLITY

::=::: PREHISPANIC
ROAD

MOUNTAIN

○ MODERN TOWN

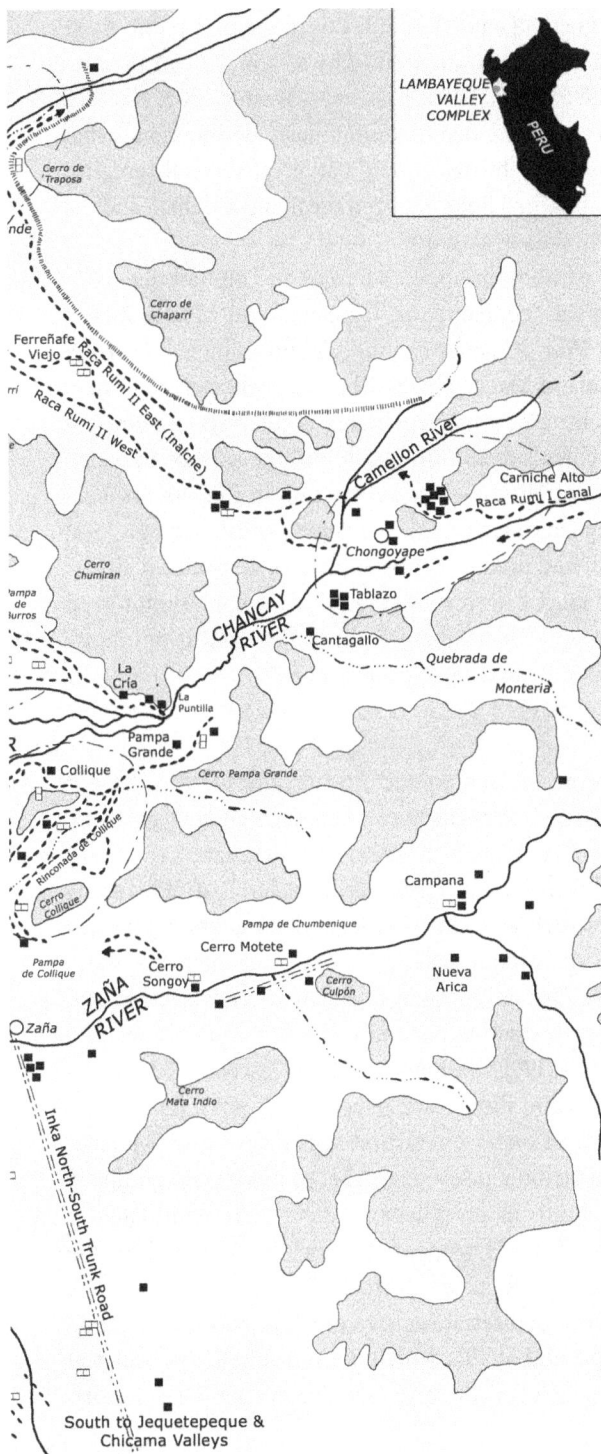

Figure 6.1. The Lambayeque Valley Complex of northern Peru, including the sites discussed in the text (figure redrawn by Haagen Klaus from Shimada [1994: figure 3.15] and based on an unpublished map by Paul Kosok in the possession of the late Richard P. Schaedel).

of an extended corpse aligned on a north-south axis in a simple pit no more than two meters in any dimension, accompanied by ceramic, metal, or camelid offerings (Donnan 1995; Klaus 2003; Klaus and Wester 2005; Narváez 1995; Shimada 2000). Recent evidence unambiguously demonstrates that practices of prolonged primary burial, manipulation of skeletal remains, and secondary burials maintained links between the living Mochica and the sacred ancestral dead (Shimada et al. n.d.).

Adherence to Mochica traditions appears unique in Lambayeque; contemporaneous burials in the Jequetepeque, Chicama, and Moche valleys to the south reflect later Wari, Cajamarca, and Chimú influence (Bernuy and Bernal 2005; Donnan and Mackey 1978). The persistence of traditional Mochica burial patterns in the Lambayeque area is inferred to represent an embodied ethnic Mochica substratum just under the surface of Sicán, Chimú, and Inka dominion. The Mochica substratum in the late pre-Hispanic Lambayeque Valley Complex signals a unique persistence of a local identity, cultural memory, and practical consciousness that outlived all imposed political systems. Practice of this identity by local peoples reinforced social boundaries and group solidarity, especially in the context of intense interaction with Sicán, Chimú, and Inka ethnic groups.

The Colonial Period to 1750

A wealth of information on Andean contact and its aftermath is present in ethnohistoric sources from the southern Inka highlands (e.g., Cobo 1653 [1990]; Guamán Poma de Ayala 1615 [1980]). Unfortunately, the north coast lacked a dedicated Spanish or indigenous chronicler. Fastidious study of nonsystematic sources such as deeds and legal documents helps shed light on the overlooked Lambayeque region. Europeans dismantled indigenous *parcialidades*, a pre-Hispanic system of dualistic sociopolitical and economic organization that extended back into the first millennium AD, if not earlier (Netherly 1984, 1990; Shimada 2001). As early as 1534, *encomenderos* confiscated lands and forced native peoples to work for Spanish profit (Mendoza 1985). *Reducciones* were instituted extensively by 1540, creating indigenous labor and tribute pools. A market economy was gradually installed. Camelids—linchpins of pre-Hispanic diet—were eliminated in favor of large-scale husbandry of European livestock, slaughtered for the production of soap and leather (Ramírez Horton 1974). Environments were degraded by desertification, deforestation, overgrazing, and introduction of water-hungry sugarcane and alfalfa, which periodically left no water at the end of the irrigation canals for Mochica farmers; during the dry years, they even lacked drinking water (Ramírez 1996: 74). Local lords were transformed into tribute collectors. Social and cosmological ties to ancestral lands were severed. The Mochica struggled to meet increasing labor demands as they sank into poverty and lost respect for their indigenous leaders, while indigenous systems of collective well-being disintegrated (Ramírez 1996: 157).

Little direct information exists regarding religion in colonial Lambayeque. It can be inferred, however, that the Catholic Church was a powerful agent of change. Oral traditions state that those who were deemed heretics were burned alive atop pre-Hispanic *huacas* (monumental adobe pyramids) at Túcume, which was decreed to be the physical gateway to hell (Heyerdahl et al. 1995: 212).

The contact-era generation in Lambayeque was likely ravaged by European pathogens. Aggregate north coast census data reveal devastating decline, from an estimated 321,000 inhabitants in 1520 to 23,578 individuals by 1630 (Cook 1981). Yet depopulation was neither universal nor synchronous. Due to its large economic promise, the large indigenous population of the Lambayeque Valley Complex declined minimally from 1570 to 1620 despite several waves of epidemics. Disease, migrations of working males, and population resettlements all had varying consequences; but by the mid-sixteenth century the Lambayeque region's unequaled economic potential and large precontact population helped initiate a demographic rebound (Keith 1976).

The Chapel of San Pedro de Mórrope

The traditional Mochica town of Mórrope is adjacent to the southern margin of the Sechura Desert in the environmentally marginal northwest corner of the Lambayeque Complex, 803 km north of Lima, Peru. Surface surveys suggest a continuous local population in the area from at least the Moche V period (AD 550–750). The town was one of the last bastions of the extinct Muchik language. Mochica surnames persist today, and the community-level ceramic craft specialization is based on thousand-year-old *paleteada* technology (Cleland and Shimada 1998).

The Chapel of San Pedro de Mórrope was founded in 1536 and features European-style external architecture (figure 6.2a, b). Internally, pre-Hispanic building styles are found in the use of adobe bricks and algarrobo (*Prosopis* sp.) tree trunks as ceiling beams and Y-shaped posts, as depicted in Moche-period fineline paintings. The adobe brick altar is a three-dimen-

Figure 6.2. The Chapel of San Pedro de Mórrope, as seen from the exterior (a) and interior (b). The chapel interior shows plastered post-and-beam construction and the three-meter-high stepped-pyramid altar in the background (photos by Cesar Maguiña/ICAM).

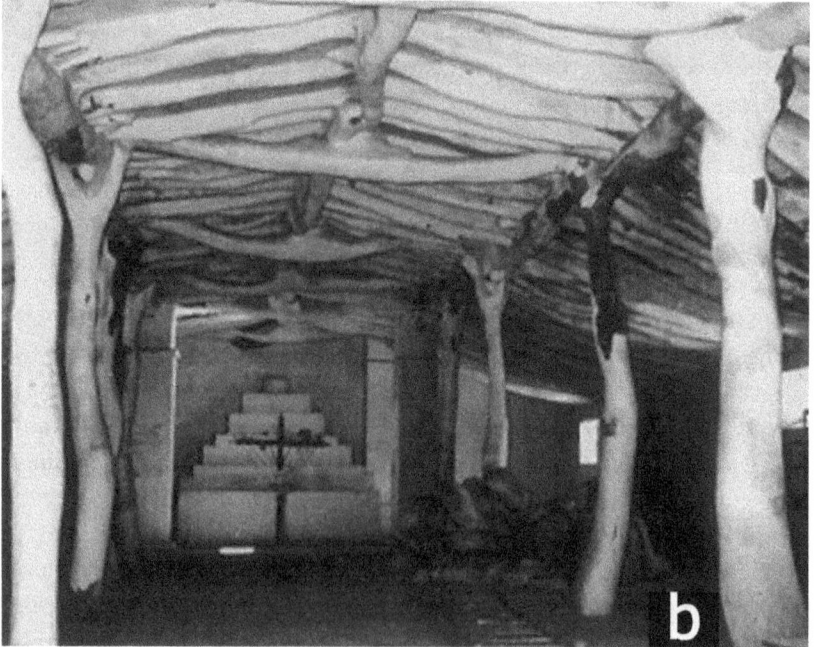

sional stepped pyramid (an icon formalized during the Moche era), probably representing cosmologically powerful mountains or *huacas*. The chapel was abandoned sometime between 1720 and 1751. Since then it slowly decayed to the point of collapse by the early twenty-first century.

The only ethnohistoric source on colonial Mórrope is an unfinished eighteenth-century manuscript by the Mórrope priest Don Justo Modesto Rubiños. Although many Spanish accounts exhibited "layer after layer of European ethnocentric veneer, bias, misdepiction, and misunderstanding of indigenous ideas and concepts" (Ramírez 1996: 152), a striking vision of colonial Mórrope still emerges from the Modesto Rubiños (1782 [1936]) manuscript. The first mass was consecrated at the chapel on June 29, 1536, and the Mochica, characterized as irrational, ignorant brutes, began to "give in to God," in the words of Modesto Rubiños. Still, conversion of the population was portrayed as a challenging task even into the seventeenth and eighteenth centuries. Many Mochica were baptized but continued traditional social and religious practices, including shamanism, practitioners of which were actively persecuted. In 1658 many religious "errors" and misinterpretations of Catholic practices were committed by even the most Latinized and "rational" Mochica. Allusions to violence against the Mochica occur throughout the document, including the killing of a local principal lord, manipulating the Mochica in power struggles between Europeans, and the threat of grave punishments for those who would not assimilate. In 1667 particularly strong punishments were said to have been inflicted on those who disregarded the ban on imbibing *chicha* (fermented maize beer) in practices that were cornerstones of identity and social solidarity. In 1780 Modesto Rubiños himself engaged in a bitter power struggle with local Mochica leaders. He used military force to imprison them in the highlands, under the pretext that they were promoting satanic dances and drunkenness.

The Spanish placed Mórrope into the encomienda of Jayanca, which lost 81% of its population from 1575 to 1602 (Cook 1981: 143–144). According to Modesto Rubiños, the size of the population in Mórrope fluctuated over the years. The population of 697 people in 1536 rose to 1,548 individuals in 1548, as Mórrope absorbed two *parcialidades* of nearby Pacora and families from Eten. By 1645 the Mórrope population fell to perhaps as few as 900 individuals but since then has steadily grown into the twenty-first century to about 4,000 people. Other passages point to volatile conflicts with neighboring communities over sparse water, land, and mining rights. Colonial Morropanos are said to have cultivated mostly maize, beans, and peppers; if this statement is accurate, it represents a major reduction in cultigen diversity from pre-Hispanic times.[1] Morropanos appear to have been exploited as draft laborers in nearby gypsum and salt mines and also played a critical role in trans–Sechura Desert transport and communication, gaining them the name "desert walkers" (Figueroa and Idrogo 2004: 131).

The propaganda-like narrative of Modesto Rubiños is highly laudatory of the priests who succeeded in bringing salvation to a land of gentiles, extinguishing superstition, ignorance, and the devil's influence in Mórrope. The Modesto Rubiños manuscript provides a European's vision of what today we would interpret as atrocities driving the suppression of the identity of the Mochica, who overall appeared progressively divorced from pre-Hispanic belief, ethnicity, and memory—conforming to what John Bodley (1999) would term "ethnocide".

As part of the Lambayeque Valley Biohistory Project directed by Klaus, archaeological excavations with Tam in 2004–2005 at the Chapel of San Pedro de Mórrope resulted in the recovery of 322 burials. Multivariate statistical comparison with late pre-Hispanic Lambayeque Valley samples (AD 900–1532; $n = 255$),[2] which represent the full range of known social variation from fishers to paramount lords, shows that the population at Mórrope experienced considerably poorer health: female fertility was depressed; oral health declined as the diet became focused on starchy carbohydrates; and the prevalence of periosteal, porotic hyperostosis, and degenerative joint disease lesions increased (figure 6.3) (Klaus 2008; Klaus and Tam 2007). Thus the biological repercussions of missionization and European colonization for the Mochica were similar to those manifested in other regions of North (Larsen 1994, 2001), Central (Storey et al. 2002), and South America (Ubelaker and Newson 2002).

At the Chapel of San Pedro de Mórrope, many of the dead were placed in extended positions either inside wooden coffins or in textile burial shrouds (figure 6.4). Coffins were adorned with copper tacks, sometimes arranged in the form of the Christian cross. Many bodies were dressed in simple European-style garb. Desiccated flowers accompanied the dead in at least 41 burials. Beyond that, grave offerings were absent. Stratigraphic position, seriation, and multiple correspondence analysis of 27 mortuary pattern variables allowed the partition of the burial sample into Early/Middle and Middle/Late Colonial phases.

At the chapel itself, 235 burials were oriented on a north-south axis, usually with the head facing south toward the altar. Traces of a red pigment or

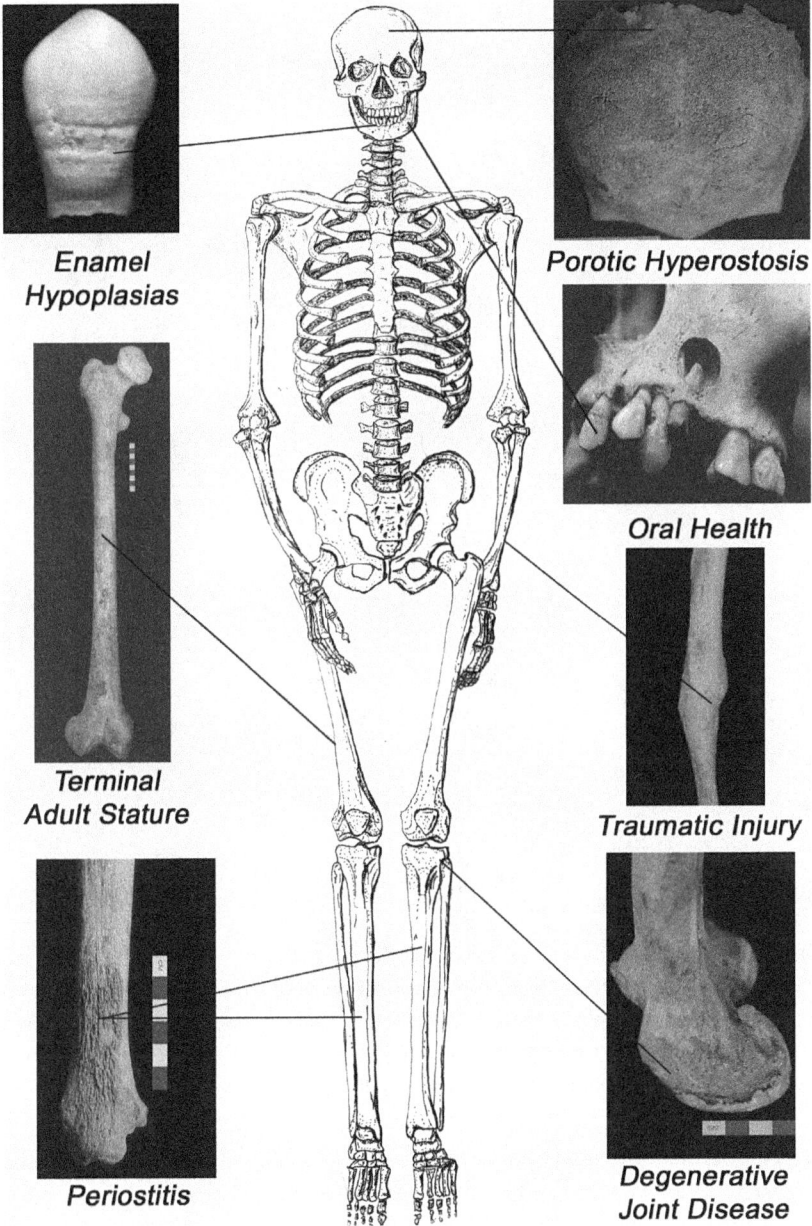

Enamel
Hypoplasias

Porotic Hyperostosis

Oral Health

Terminal
Adult Stature

Traumatic Injury

Periostitis

Degenerative
Joint Disease

Figure 6.3. Characteristics and pathological conditions used to establish a multiple-attribute health profile of the pre- and postcontact Lambayeque Valley Mochica populations.

Figure 6.4. The Chapel of San Pedro de Mórrope, Unit 3, Level 8 (170–189.9 cm below datum). The 18 Middle to Late Colonial interments here are representative of basic Mórrope mortuary patterns, which superficially resemble European funerary practice but simultaneously encode pre-Hispanic identity (photo by Haagen Klaus).

red textile were documented covering the faces of at least 37 individuals. In 3 cases, evidence of fires adjacent to the burial pit was observed.

Careful attention to burial taphonomy (see Duday 2006; Roksandic 2002) reveals that at least 114 individuals in the Mórrope cemetery were buried shortly after death and the corpse decayed undisturbed. Evidence of mortuary programs linking the living and the dead was observed elsewhere, discussed in depth by Klaus and Tam (n.d.). Burial was prolonged, perhaps a month or longer in at least 24 cases, evidenced by masses of empty fly puparia and remains of beetles that are predacious on fly eggs and maggots inside textile bundles, coffins, bony cavities, and desiccated brain tissue.

Postinterment removal of body parts was observed among 133 burials. Some altered burials clearly resulted from subsequent inhumations, but a substantial number correspond to intentional opening of a grave or coffin for the removal of human remains, specifically skulls and long bones (figure 6.5). Burial U705-9 was one of several cases with the coffin lid missing and the child's head, right hand, legs, and feet removed. Other burials were apparently exhumed and redeposited after coffin lids were removed and bones inside manipulated; in some cases, additional long bones and crania were placed inside the coffin with the primary burial. A small number of coffins were empty, though traces of an original occupant remained.

The 53 secondary burials that were excavated were compositionally dominated by the crania and long bones of at least 327 individuals. Secondary burial of isolated bones was also very common and included 2,569 long bones and several thousand more skeletal elements yet to be studied.

Pre-Hispanic Rituals and Ethnogenesis in Colonial Mórrope

Biological Hybridization and Ritual Syncretism in Colonial Mórrope

Available ethnohistoric data indicate that the Mórrope Mochica endured nothing less than postcontact cultural disaster marked by violence, conversion, social restructuring, powerlessness, and poverty. As health stress reached chronic levels, socially destabilizing human pain, suffering, and psychological stress ensued. Ethnohistoric and biological data signal biosocial collapse, which is traditionally conceived as a condition of ethnocide. Were these consequences indeed experienced by the Mochica? Biological interaction patterns and mortuary behaviors, we think, hold the key to responding to the question of indigenous identity in colonial Mórrope.

Our initial study of Mochica population structures and biological interactions before and after Spanish arrival is currently tentative but potentially very informative. Inherited tooth sizes were subjected to an R-matrix analysis and a conservative calculation of minimum F_{ST}, which serves to estimate population genetic variance (Relethford and Blangero 1990; Stojanowski 2004, 2005). The results show a precipitous drop in overall genetic diversity from the late pre-Hispanic to the postcontact era from 0.041 to 0.009, respectively. The postcontact group is homogenous and has little-to-

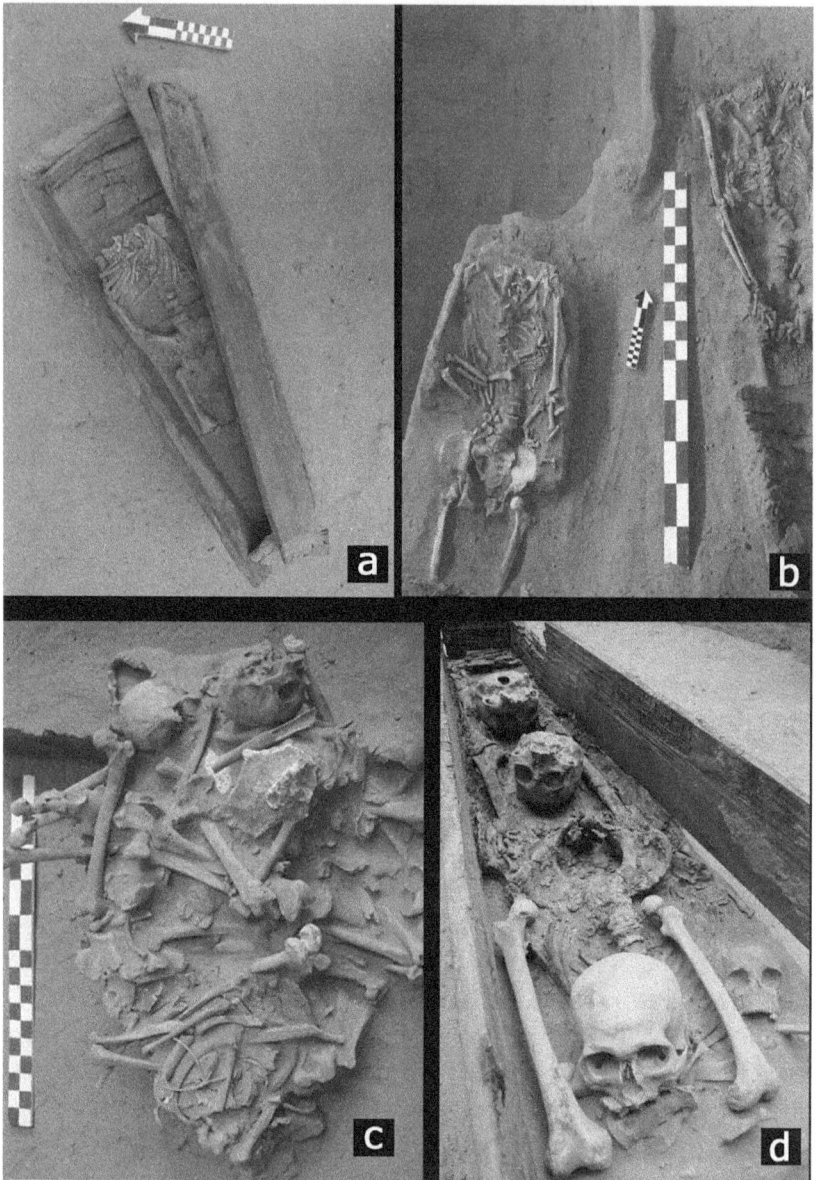

Figure 6.5. Altered burials from the Chapel of San Pedro de Mórrope, including (*a*) a subadult whose coffin had been opened (coffin lid missing) and the head, leg long bones, and feet removed (Burial U705-9); (*b*) two individuals whose skulls had been removed postdepositionally (Burials U405-29 and U405-30); and (*c*) a typical large secondary burial, consisting mostly of long bones and crania (Burial U505-1) (photos by Haagen Klaus); in addition, Burial U705-2 (*d*) originally contained the skeleton of an elderly adult female. The coffin was subsequently opened (coffin lid missing) and the skull and upper limbs removed. Then the crania of four individuals were added, along with an additional left hand and other long bones, including two femora from distinct individuals (photo by Haagen Klaus).

no biological structure. One explanation could be the stochastic effects of random genetic drift; loss of biological variation was due to catastrophic depopulation, fugitivism, and outmigration of workers, which removed a large quantity of alleles from the indigenous gene pool.

Historic records, however, report an average of only 39.9% depopulation in the Lambayeque region from 1575 to 1602. While demographic decline no doubt contributed to changes in F_{ST}, another process was likely transpiring. Estimations of patterns of differential gene flow from the R-matrix shed further light. The Early/Middle Colonial Mochica population experienced an increase in relative gene flow at a level nearly doubled from the late pre-Hispanic period. By the Middle/Late Colonial period, the direction of gene flow reversed to an extreme degree, falling nearly 190%.

These observations very closely parallel those by Stojanowski (2004, 2005, this volume) in Spanish Florida. Depopulation, broadly similar Spanish economic policies, and indigenous biocultural adaptations resulted in a similar outcome. In practical terms, these data reveal that native groups reconfigured and expanded mating networks as traditional systems became increasingly dysfunctional. Increased gene flow between the survivors of contact and their descendants created a far more homogenous population. This probably happened quickly in Mórrope, within the first 100 years or so following contact. By the Middle/Late Colonial phase, aggregation and biological hybridization was completed. Gene flow signatures subsided. *Reducciones* provided the physical setting for this to happen, as they aggregated people to an unprecedented degree. But the underlying key to this process of hybridization would have necessitated breaking down social barriers to widespread gene flow: namely, those aspects of identity associated with membership in relatively endogamous *parcialidad* hierarchies. Also, initial kinship analysis of intraindividual dental traits reveals not a single discernible spatial relationship between phenetically similar individuals buried in Mórrope, which contrasts with the often sharply defined bio-spatial nature of late pre-Hispanic cemetery planning. This cultural discontinuity probably reflects boundary deconstruction as well. Such maintenance of social group distinctions no longer held much meaning (Klaus 2008).

Mortuary patterns hold the other key to understanding ethnogenesis in colonial Mórrope. In interpreting the significance of mortuary patterns, rigor and caution must be used. Prudent and thoughtful evaluation must validate analogies and inferences based on archaeological or ethnohistoric data, especially those derived from outside the north coast of Peru. An apparent match between archaeological and historical information must be framed in terms of an in-depth understanding of cultural context and other lines of evidence, guarding against disjunctions between form and meaning, potential homologies, or convergences (Kubler 1970; Shimada et al. 2004).

While superficially Catholic in appearance, the burials at Mórrope involve multiple symbolic subtexts representing practice and reproduction of pre-Hispanic rituals just under the surface. These features are distributed in both the Early/Middle and Middle/Late Colonial mortuary record but are most strongly concentrated in the Middle/Late temporal component.

First, extended bodies seem European in style, but the burials (and even the chapel) are oriented on a north-south axis, which was the common orientation of pre-Hispanic Mochica burials. Body orientation at Mórrope contrasts with Christian doctrine, which held that the dead were to be placed on an east-west axis so they that would be resurrected to face the returning Christ. While its more emic meaning is difficult to access, the pre-Hispanic Mochica *axis mundi* was reproduced at Mórrope.

Second, a number of children and adults were adorned with a red textile covering the face, which we propose directly parallels pre-Hispanic ritual in which cinnabar was painted on the face or body, perhaps symbolizing life-force as embodied in blood. At Mórrope the dye appears organic in nature and lacks the brilliance of mercuric hematite, suggestive of a colonial-era attempt to reproduce this practice while lacking access to cinnabar.

Third, burning adjacent to Early Colonial graves mirrors the pre-Hispanic practice, where ritual fires were set within or near a burial pit (Shimada et al. n.d.).

Fourth, prolonged and altered burials in colonial Mórrope appear virtually identical to distinct precontact rituals that linked the living and the dead (Shimada et al. n.d). Maggot infestation of at least 21 individuals mirrors a pre-Hispanic concept of soul-transfer. Earlier Moche iconography suggests that a fly may have symbolized the "soul" of the deceased (Bourget 2006), echoing the Quechua conception present in colonial Huarochirí folklore of the Lima highlands. As in multiple pre-Hispanic examples, permitting the body to putrefy and decompose preceding burial may have served to liberate the spiritual component of a person from the corporeal skin and bones (Salomon 1995: 330).

The many altered and secondary burials at Mórrope share a highly repetitive pattern with their local pre-Hispanic analogs: predominant manipulation and reburial of skulls and long bones illustrated by univariate distributions and multiple correspondence analysis (Klaus 2008). Focus on heads and long bones appears to reflect a consistent, conscious cultural choice regarding selective reburial, but why?

It is often suggested that living-dead interactions in the Andes represent ancestor worship based on an ethnohistoric, Inka-centric model, while the word "ancestor" is an overused and abused gloss for "the dead." Multiple lines of archaeological and skeletal evidence argue against ancestor worship in Mórrope (Klaus and Tam n.d.). Manipulation of the dead may reflect instead an indigenous concept of fertility. On the pre-Hispanic north coast, the central power of the dead likely involved their influence on the fertility of the living world—a theme widely shared in the Andes and cross-culturally (Bloch and Parry 1982). One of the most persistent and widespread aspects of Andean cosmology is a vegetal metaphor of the dead. After the fleshy body rots away, bones are likened to the seeds that fall from a dying plant, which, when treated properly, are the source of new life. Remains of adults may have embodied the fertility of the immutable existence of very old beings (Salomon 1995: 328). Conversely, children may not have been perceived as human in the Andes and were instead linked to the mountain domain from which their spirits originated (Sillar 1994). Observed manipulation of dead children may have accessed their liminality, bridging supernatural and earthly domains to serve as a potent conduit for the living to appeal for water, fertile fields, and reproduction of a threatened social cosmos as well.

Exhuming skeletons apparently harvested raw materials, and subsequent secondary burials may have completed a process embodying a metaphor of seed planting in an attempt to direct the power of the dead to ensure communal well-being. Comparing the number of elements reinterred under the chapel floor versus the minimum number of individuals reburied, an astounding 94.6% of human remains (mostly non–long bones) are simply missing. Bones may have been removed from the chapel, curated, enshrined, or buried within a household or workshop. Perhaps bones were taken into agricultural fields or other locations and quite literally planted. Beyond the interpretation of flowers (the only consistently used grave good) placed with the dead as tokens of affection or mourning, we consider the possibility that vegetal metaphors were possibly tied to flower offerings as well.

Identity Manipulation and Ethnogenesis in Colonial Peru

Ethnogenesis in Mórrope appears as a two-stage process: the first manifested in the biological sphere, and the second expressed in ritual domains.

First, the Mochica reconfigured biological identity on the level of the self and the group. In Spanish Florida and Peru, ethnogenesis resulted from a constellation of indigenous survival strategies where so-called hybridized group coalitions emerged as a response to demographic and biological threat. Group membership was widened to the lowest acceptable common denominator to ensure the stability, if not very existence, of the group. As Mórrope tooth size and nonmetric dental traits indicate, Mochica perception of ethnicity did not extend widely to Europeans or African slaves, if at all (Klaus 2008). Another factor may have been native perceptions of group identity following imposition of the concept of el indio. This Spanish vision of a single "Indian" identity was used by indigenous southern Andeans as a self-descriptive term as early as the 1560s and no doubt aided deconstruction of traditional identity boundaries and group hybridization.

Once this primarily biological phase was over by the close of the Early/ Middle Colonial sequence, a second stage of ethnogenesis took over, involving the creation and practice of syncretic Euro-Andean cultural patterns found mostly in the Middle/Late Colonial mortuary record. At first glance, it would be easy to think that a Christianized identity was manifested in Mórrope. The Mochica buried their dead in sanctified ground, often in European-style coffins sometimes decorated with the Christian cross. The loss of traditional grave elaboration and status symbolism likewise indicates symbolic if not practical denial of pre-Hispanic social structures and ranking systems. Considering the use of clothing as a tool of identity, European accessories such as shoes, buttons, and textiles of nontraditional manufacture in many Middle/Late Colonial Mórrope burials hold important implications. At least in death, a shifting identity in a Latinized direction may have preceded enforcement of dress laws in the 1780s.[3]

More deeply, reproduction of precontact rituals inside the Chapel of San Pedro de Mórrope illustrates modification, rather than annihilation, of a conscious indigenous identity intimately linked to precontact rituals and their meanings. The deeply rooted and conservative pre-Hispanic Mochica substratum, which weathered Moche and Sicán collapses and conquest by the Chimú and Inka, remained viable if perhaps driven somewhat "underground" following contact.

Robert Hertz (1960), Peter Metcalf and Richard Huntington (1991), Arnold van Gennep (1960), and others remark on how human remains can be divorced from the biological dimensions of death and become a locus of social expression. As such, Mórrope death rituals illuminate aspects of the colonial social order, especially when viewed in terms of symbolism and agency. Burial on the traditional *axis mundi* and secondary burial would have embodied social memory, values, and ideology while conveying an emotional force stimulating the reflexive construction of group identity (Bawden 2005). Secondary burials, such as those carried out in Mórrope, have long been interpreted as an assertion of collectivity and cohesion (Hertz 1960). As Debra Gold (2000: 196) points out, the kinds of secondary burial at Mórrope would have required the collaboration and collective work of a sizable number of living individuals, bringing them in contact with the dead. Living-dead interactions at Mórrope may echo a colonial Mochica practical consciousness (Giddens 1979), generating an intense experience of *communitas*.

Mochica agency and creativity are the most probable source behind the reproduction of 1,500-year-old traditions inside Christian rituals. On deep levels, the basic ritual patterns practiced at Mórrope would have been set within an isochrestic plane where stylistic variation is set by enculturation and *habitus*; but the active use of these symbols in this colonial context appears very much emblemic, as burials functioned as consciously manufactured socio-ethnic markers (see Reycraft 2005 for other Andean examples). The kind of syncretism seen here may actually have been conditioned by the early Mórrope priests themselves: some of them were Augustinians, who, in their quest for converts, were more open to a spiritual compromise and tolerant of syncretism. The emic meanings of a red textile, for instance, may not have been apparent to a priest but were shared by Mochica funerary participants. The Mochica may also have carried out ancestral traditions surreptitiously when the priest ministered to nearby Pacora. Overall, a hybrid ritual complex was entrenched by the eighteenth century. Perhaps these were among the "errors" and "misunderstandings" that Modesto Rubiños (1782 [1936]) lamented.

The people of Mórrope confronted a profound social and structural crisis, using their changing identity as a tool if not an outright weapon of ideological conflict. Mórrope mortuary patterns may be seen as a form of symbolic resistance against colonization, while simultaneously representing both concrete and symbolic success against involuntary religious conversion and extirpation. It is also constructive to consider Elizabeth Graham's (1998) proactive, rather than reactive, stance; postcontact adaptive responses were not simply to thwart domination but were in parallel with creative attempts to understand and gain conceptual mastery over radical change. As Graham (1998: 29) writes, "[r]esistance and protest occur alongside active reexamination of former values, together with the development of new concepts about the world that indeed receive European input, but are the product of indigenous minds."

Conclusions

This bioarchaeological study of colonial north-coast Peru indicates that when the Chapel of San Pedro de Mórrope was abandoned, sometime during the early to mid-eighteenth century, a new kind of person and culture that had never existed before—the colonial Mochica—had come into existence. Highly negative Spanish socioeconomic transformations were key factors leading to indigenous experiences of biological stress. Yet the Mochica neither demographically imploded nor culturally collapsed. Instead, ethnogenesis was the result, involving a rapid indigenous biological hybridization that concluded with a cultural hybridization, leaving perceptible traces in syncretic mortuary behaviors. This process involved pre-Hispanic identities engaged in a dialectical struggle with the colonial reality, while unprecedented health burdens no doubt threatened to destabilize an emergent regional Mochica biocultural coalition. These responses helped in successfully buffering against the possibility of unrecoverable biocultural trauma in the Lambayeque Valley. In a hemispheric comparison, ethnogenesis for the natives of Spanish Florida was not fully realized by the time regional conflict destroyed the mission system by 1706 (Stojanowski 2005). In Peru there was no arrest of indigenous ethnogenesis by global power struggles, and a more reintegrated hybrid configuration could develop following the conclusion of native biological homogenization.

Our methodological intention has been to demonstrate that the integration of ethnohistoric, skeletal biological, and archaeological data is vital to frame the bioarchaeology of identity (or any social phenomenon for that matter) in terms of multiple variables interacting with local factors rooted in a historic context (Palkovich 1996). Synthesis of dental phenetic and burial patterns opened a remarkably different view and a more holistic understanding of postcontact identity than what ethnohistoric and skeletal indicators alone suggest. Ethnohistory and bioarchaeology establish a context in which to detect and perceive the broader developments of biocultural adaptation and ethnogenesis. Considering the widespread lack of mortuary-bioarchaeology integration, wider debate and critical examination of the concept of bioarchaeology are long overdue, as is the active practice of Jane Buikstra's (1977) model of an integrative bioarchaeology.

Life for the Mochica in twenty-first-century Mórrope continues. Today mourners bring the dead to the Church of San Pedro. Following a short service, the coffin is taken to the modern cemetery and sealed in a crypt. At some point between the 1750s and the present day, colonial Mochica burial traditions largely seem to have faded into the past. But ethnogenesis continues. Over the past 30 years archaeological study of the Lambayeque region's history in particular has contributed to a reawakening and revitalization of local perceptions, questions, and embodiments of Mochica identity. The early twenty-first century will no doubt be seen as a time when the birth of yet another new Mochica identity was on the horizon.

Acknowledgments

The Wenner-Gren Foundation (Dissertation Fieldwork Grant 7302), the Tinker Foundation, and The Ohio State University's Office of International Affairs, Center for Latin American Studies, and Department of Anthropology generously funded this research from 2004 to 2006. Cesar Maguiña of ICAM served as co-director of the 2005 field season. We thank Rosabella Alvarez-Calderón, Jorge Centurión, Victor Curay, Dr. Carlos Elera, Marco Fernández, Gabriella Jakubowska, Juan Martínez, Emily Middleton, Analise Polsky, Raul Saavedra, Izumi Shimada, Carlos Wester, the people of Mórrope, and our team of undergraduate students from the Universidad Nacional de Trujillo for their many contributions. The Museo Nacional Sicán provided essential laboratory space and logistics. Clark Larsen, Paul Sciulli, Sam Stout, Daniel Temple, and the editors offered stimulating editorial comments. Any shortcomings are entirely our own.

Notes

1. Planned stable isotope studies in the near future will attempt to characterize overall dietary changes more comprehensively.

2. Sites include Sicán (Farnum 2002), Huaca del Pueblo Batán Grande (Farnum 2002), Illimo (Klaus, Fernandez, et al. 2004), Huaca Sialupe (Klaus 2003), Cerro Cerrillos (Klaus, Centurión, and Curo 2004), Huaca Cao Viejo, El Brujo Complex (Farnum 2002), Cascajales (Klaus and Wester 2005), Úcupe (Klaus and Wester 2005), La Caleta de San José (Klaus and Wester 2005), and Túcume (Toyne 2002).

3. Historic interrelationships among dress, economic activity, and identity in the region await fuller study. Photographs taken by the German ethnologist Heinrich Brüning of Lambayeque peoples (including Morropanos) in the late nineteenth and early twentieth century generally portray men in Latin dress, whereas women and children are frequently depicted wearing more traditional garb, especially in household settings.

References Cited

Baker, Brenda J., and Lisa Kealhofer, editors. 1996. *Bioarchaeology of Native American Adaptation in the Spanish Borderlands*. Gainesville: University Presses of Florida.

Barth, Fredrik. 1969. "Introduction." In *Ethnic Groups and Boundaries: The Social Organization of Culture Difference*, edited by F. Barth, 9–38. Boston: Little, Brown and Company.

Bawden, Garth. 1996. *The Moche*. Cambridge, Mass.: Blackwell.

———. 2001. "The Symbols of Late Moche Social Transformation." In *Moche Art and Archaeology in Ancient Peru*, edited by J. Pillsbury, 285–305. New Haven: Yale University Press.

———. 2005. "Ethnogenesis at Galindo, Peru." In *Us and Them: Archaeology and Ethnicity in the Andes*, edited by R. M. Reycraft, 12–33. Los Angeles: Cotsen Institute of Archaeology, University of California, Los Angeles.

Bentley, G. Carter. 1987. "Ethnicity and Practice." *Comparative Studies in Society and History* 29: 24–55.

Bernuy, Katiusha, and Vanessa Bernal. 2005. "Influencia Cajamarca en los rituales funerarios del Periodo Transicional en San José de Moro." *Corriente Arqueológica* 1: 61–77.

Bloch, Maurice, and Jonathan Parry, editors. 1982. *Death and the Regeneration of Life*. Cambridge: Cambridge University Press.

Bodley, John H. 1999. *Victims of Progress*. Mountain View, Calif.: Mayfield.

Bourdieu, Pierre. 1977. *Outline of a Theory of Practice*. Cambridge: Cambridge University Press.

Bourget, Steve. 2006. *Sex, Death and Sacrifice in Moche Religion and Visual Culture*. Austin: University of Texas Press.

Buikstra, Jane E. 1977. "Biocultural Dimensions of Archaeological Study: A Regional Perspective." In *Biocultural Adaptation in Prehistoric America*, edited by R. L. Blakely, 67–84. Proceedings of the Southern Anthropological Society 11. Athens: University of Georgia Press.

Cleland, Kate M., and Izumi Shimada. 1998. "Paleteada Potters: Technology, Production Sphere, and Sub-Culture in Ancient Peru." In *Andean Ceramics: Technology, Organization and Approaches*, edited by I. Shimada, 111–150. MASCA Research Papers in Science and Archaeology. Philadelphia: University of Pennsylvania Museum of Archaeology and Anthropology.

Cobo, Bernabé. 1653 [1990]. *Religion and Customs*. Translated and edited by R. Hamilton. Austin: University of Texas Press.

Cohen, Abner. 1974. "Introduction: The Lesson of Ethnicity." In *Urban Ethnicity*, edited by A. Cohen, ix–xxiii. London: Tavistock.

Cook, Nobel David. 1981. *Demographic Collapse: Indian Peru, 1520–1620*. Cambridge: Cambridge University Press.

Crosby, Alfred W. 1972. *The Columbian Exchange: Biological and Cultural Consequences of 1492*. Westport, Conn.: Greenwood Publishing.

Donnan, Christopher B. 1995. "Moche Funerary Practices." In *Tombs for the Living:*

Andean Mortuary Practices, edited by T. D. Dillehay, 111–159. Washington, D.C.: Dumbarton Oaks.

Donnan, Christopher B., and Carol J. Mackey. 1978. *Ancient Burial Patterns of the Moche Valley, Peru*. Austin: University of Texas Press.

Duday, Henri. 2006. "L'archéothanatologie ou l'archéologie de la mort (Archeothanatology or the Archaeology of Death)." In *The Social Archaeology of Funerary Remains*, edited by R. Gowland and C. Knüsel, 30–56. Cambridge: Oxbow Books.

Elera Arévalo, Carlos G. 1998. "The Puémape Site and the Cupisnique Culture: A Case Study on the Origins and Development of Complex Societies in the Central Andes, Perú." Doctoral dissertation, University of Calgary, Department of Archaeology.

Farnum, Julie F. 2002. "Biological Consequences of Social Inequalities in Prehistoric Peru." Doctoral dissertation, University of Missouri, Department of Anthropology.

Figueroa, Guillermo, and Ninfa Idrogo. 2004. *Lambayeque en el Perú colonial*. Chiclayo, Peru: Ingenium.

Giddens, Anthony. 1979. *Central Problems in Social Theory: Action, Structure and Contradiction in Social Analysis*. Berkeley: University of California Press.

Gold, Debra. 2000. "'Utmost Confusion' Reconsidered: Bioarchaeology and Secondary Burial in Late Prehistoric Virginia." In *Bioarchaeological Studies of Life in the Age of Agriculture: A View from the Southeast*, edited by P. M. Lambert, 195–218. Tuscaloosa: University of Alabama Press.

Goldstein, Lynne. 2006. "Mortuary Analysis and Bioarchaeology." In *Bioarchaeology: The Contextual Analysis of Human Remains*, edited by J. E. Buikstra and L. A. Beck, 375–387. New York: Academic Press.

Gowland, Rebecca, and Christopher Knüsel. 2006. "Introduction." In *Social Archaeology of Funerary Remains*, edited by R. Gowland and C. Knüsel, ix–xiv. Oxford: Oxbow Books.

Graham, Elizabeth. 1998. "Mission Archaeology." *Annual Review of Anthropology* 27: 25–62.

Guamán Poma de Ayala, Felipe. 1615 [1980]. *El primer nueva corónica y buen gobierno*. Edited by J. V. Murra and R. Adorno. 3 vols. Mexico City: Siglo XXI.

Haldane, John B. S. 2002. "Environment and History." In *The Evolution and Genetics of Latin American Populations*, edited by F. M. Salzano and M. Cátira Bortolini, 23–54. Cambridge: Cambridge University Press.

Hertz, Robert. 1960. *Death and the Right Hand*. Translated by Rodney Needham and Claudia Needham. Glencoe, Ill.: Free Press.

Heyerdahl, Thor, Daniel H. Sandweiss, and Alfredo Narváez. 1995. *Pyramids of Túcume: The Quest for Peru's Forgotten City*. London: Thames and Hudson.

Jones, Siân. 1997. *The Archaeology of Ethnicity: Constructing Identities in the Past and Present*. London and New York: Routledge.

Keith, Robert G. 1976. *Conquest and Agrarian Change: The Emergence of the Hacienda System on the Peruvian Coast*. Cambridge, Mass.: Harvard University Press.

Klaus, Haagen D. 2003. "Life and Death at Huaca Sialupe: The Mortuary Archaeology of a Middle Sicán Community, North Coast of Peru." Master's thesis, Southern Illinois University–Carbondale.

———. 2005. Lambayeque Valley Biohistory Project unpublished field notes, 2005 season.

———. 2008. "Out of Light Came Darkness: Bioarchaeology of Mortuary Ritual, Health, and Ethnogenesis, North Coast of Peru, A.D. 900–1750." Doctoral dissertation, Ohio State University, Department of Anthropology.

Klaus, Haagen D., Jorge Centurión, and Manuel Curo. 2004. "New Evidence of Human Sacrifice on the North Coast of Peru: Middle Sicán Ritual Killing in the Lambayeque Valley." *American Journal of Physical Anthropology Supplement* 38: 127.

Klaus, Haagen D., Marco Fernández, Juan Martínez, and Carlos Wester. 2004. "The Cemetery of El Arenal and the Warrior of Illimo: A Biocultural Study of Middle Sicán Social Organization, North Coast of Peru." Paper presented at the 69th Annual Meeting of the Society for American Archaeology, Montreal, Quebec, Canada.

Klaus, Haagen D., and Manuel E. Tam. 2007. "Bioarchaeology of European Conquest in Peru: Health, Identity, and Ethnogenesis in the Lambayeque Valley (AD 1536–1751)." Paper presented at the 35th Annual Midwest Conference on Andean and Amazonian Archaeology and Ethnohistory, Carbondale, Illinois.

———. n.d. "*Requiem Aeternum, dona eis, Domine*: The Dead, the Living, and Pre-Hispanic Burial Ritual in Colonial Mórrope, North Coast of Peru." In *Between the Living and the Dead: Cross-Disciplinary and Diachronic Perspectives, Volume 1: Andes*, edited by I. Shimada and J. Fitzsimmons. Tucson: University of Arizona Press.

Klaus, Haagen D., and Carlos Wester. 2005. "Health, Burial, and Social Structure at Úcupe: A New Perspective on the Chimú of Ancient Northern Coastal Peru." *American Journal of Physical Anthropology Supplement* 40: 131.

Kubler, George. 1970. "Period, Style, and Meaning in Ancient American Art." *New Literary History* 1: 127–144.

Larsen, Clark Spencer. 1994. "In the Wake of Columbus: Native Population Biology in the Postcontact Americas." *Yearbook of Physical Anthropology* 37: 109–154.

———, editor. 2001. *Bioarchaeology of Spanish Florida: The Impact of Colonialism*. Gainesville: University Press of Florida.

Larsen, Clark Spencer, and George R. Milner, editors. 1994. *In the Wake of Contact: Biological Responses to Conquest*. New York: Wiley-Liss.

Mendoza, Eric, editor. 1985. *Presencia histórica de Lambayeque*. Chiclayo, Peru: Sociedad de Investigación de la Ciencia, Cultura y Arte Norteño.

Metcalf, Peter, and Richard Huntington. 1991. *Celebrations of Death: The Anthropology of Mortuary Ritual*. 2nd ed. Cambridge: Cambridge University Press.

Millaire, Jean-François. 2002. *Moche Burial Patterns: An Investigation into Prehispanic Social Structure*. BAR International Series 1066. Oxford: Archaeopress.

Modesto Rubiños, Justo. 1782 [1936]. "Noticia previa por el Lic. D. Justo Modesto Rubiños, y Andrade, cura de Mórrope año de 1782." *Revista Histórica* 10: 291–363.

Narváez, Alfredo. 1995. "La muerte en el antiguo Túcume: El Cementerio Sur y la Huaca Facho." In *Túcume*, by T. Heyerdahl, D. Sandweiss, A. Narváez, and L. Millones, 207–219. Lima: Banco de Crédito.

Netherly, Patricia. 1984. "The Management of Late Andean Irrigation Systems on the North Coast of Peru." *American Antiquity* 49: 227–254.

———. 1990. "Out of Many, One: The Organization and Rule of the North Coast Polities." In *The Northern Dynasties: Kingship and Statecraft in Chimor*, edited by M. E. Moseley and A. Cordy-Collins, 461–505. Washington, D.C.: Dumbarton Oaks.

Palkovich, Anne M. 1996. "Historic Depopulation in the American Southwest: Issues of Interpretation and Context-Embedded Analyses." In *Bioarchaeology of Native American Adaptation in the Spanish Borderlands*, edited by B. J. Baker and L. Kealhofer, 179–197. Gainesville: University Press of Florida.

Pillsbury, Joanne, editor. 2001. *Moche Art and Archaeology in Ancient Peru*. New Haven: Yale University Press.

Ramírez, Susan. 1996. *The World Upside Down: Cross-Cultural Contact and Conflict in Sixteenth-Century Peru*. Stanford: Stanford University Press.

Ramírez Horton, Susan. 1974. *The Sugar Estates of the Lambayeque Valley, 1670–1800: A Contribution to Peruvian Agrarian History*. Madison: University of Wisconsin Land Tenure Center.

Relethford, John H., and John Blangero. 1990. "Detection of Differential Gene Flow from Patterns of Quantitative Variation." *Human Biology* 62: 5–25.

Reycraft, Richard M., editor. 2005. *Us and Them: Archaeology and Ethnicity in the Andes*. Los Angeles: Cotsen Institute of Archaeology, University of California, Los Angeles.

Roksandic, Mirjana. 2002. "Position of Skeletal Remains as a Key to Understanding Mortuary Behavior." In *Advances in Forensic Taphonomy: Method, Theory, and Archaeological Perspectives*, edited by W. D. Haglund and M. H. Sorg, 99–117. Boca Raton, Fla.: CRC Press.

Salomon, Frank. 1995. "'The Beautiful Grandparents': Andean Ancestor Shrines and Mortuary Ritual as Seen through Colonial Records." In *Tombs for the Living: Andean Mortuary Practices*, edited by T. D. Dillehay, 315–353. Washington, D.C.: Dumbarton Oaks.

Schwartz, Glenn M., and John J. Nichols, editors. 2006. *After Collapse: The Regeneration of Complex Societies*. Tucson: University of Arizona Press.

Shimada, Izumi. 1994. *Pampa Grande and the Mochica Culture*. Austin: University of Texas Press.

———. 2000. "The Late Prehispanic Coastal Societies." In *The Inca World: The Development of Pre-Columbian Peru, AD 1000–1534*, edited by L. Laurencich Minelli, 49–110. Norman: University of Oklahoma Press.

———. 2001. "Late Urban Moche Craft Production: A First Approximation." In *Moche Art and Archaeology in Ancient Peru*, edited by J. Pillsbury, 177–205. New Haven: Yale University Press.

Shimada, Izumi, Haagen D. Klaus, Rafael Segura, and Go Matsumoto. n.d. "Living with the Dead: Conception and Treatment of the Dead on the Central and North Coast of Peru." In *Between the Living and the Dead: Cross-Disciplinary and Diachronic Perspectives, Volume 1: Andes*, edited by I. Shimada and J. Fitzsimmons. Tucson: University of Arizona Press.

Shimada, Izumi, Ken-ichi Shinoda, Julie F. Farnum, Robert Corruccini, and Hirokatsu Watanabe. 2004. "An Integrated Analysis of Pre-Hispanic Mortuary Patterns: A Middle Sicán Case Study." *Current Anthropology* 45: 369–402.

Sillar, Bill. 1994. "Playing with God: Cultural Perceptions of Children, Play, and Miniatures in the Andes." *Archaeological Review from Cambridge* 13: 47–64.

Stojanowski, Christopher M. 2004. "Population History of Native Groups in Pre- and Postcontact Spanish Florida: Aggregation, Gene Flow, and Genetic Drift on the Southeastern U.S. Atlantic Coast." *American Journal of Physical Anthropology* 123: 316–332.

———. 2005. "The Bioarchaeology of Identity in Spanish Colonial Florida: Social and Evolutionary Transformation before, during, and after Demographic Collapse." *American Anthropologist* 107: 417–431.

Storey, Rebecca, Lourdes Marquez Morfin, and Vernon Smith. 2002. "Social Disruption and the Maya Civilization of Mesoamerica: A Study of Health and Economy of the Last Thousand Years." In *The Backbone of History: Health and Nutrition in the Western Hemisphere*, edited by R. H. Steckel and J. C. Rose, 283–306. Cambridge: Cambridge University Press.

Toyne, Marla. 2002. "Tales Woven in Their Bones: The Osteological Examination of the Human Skeletal Remains from the Stone Temple at Túcume, Peru." Master's thesis, University of Western Ontario, Department of Anthropology.

Ubelaker, Douglas H., and Linda A. Newson. 2002. "Patterns of Health and Nutrition in Prehistoric and Historic Ecuador." In *The Backbone of History: Health and Nutrition in the Western Hemisphere*, edited by R. H. Steckel and J. C. Rose, 343–375. Cambridge: Cambridge University Press.

van Gennep, Arnold. 1960. *The Rites of Passage*. Translated by M. Vicedom and S. Kimball. Chicago: University of Chicago Press.

Verano, John W., and Douglas H. Ubelaker, editors. 1992. *Disease and Demography in the Americas*. Washington, D.C.: Smithsonian Institution Press.

PART II

Identity Formation and Manipulation at the Level of the Individual

7

Cultural Embodiment and the Enigmatic Identity of the Lovers from Lamanai

CHRISTINE D. WHITE, FRED J. LONGSTAFFE, DAVID M. PENDERGAST, AND JAY MAXWELL

Excavations approximately twenty years ago at Lamanai, Belize, uncovered a burial that we believe is still unique in the Maya world. The burial, of three individuals (figure 7.1), lay in a grave dug as part of the rebuilding of the stair of Structure N11-5 during the Late Postclassic period, probably circa AD 1450–1500. At the time, David Pendergast (1989: 1) described it as "the most interesting and enigmatic burial of the 900 or so [he] had recorded," and none of the approximately 200 that he recorded afterward altered this description. The unique quality of the burial lies in an expression of affection not yet found elsewhere among either ancient Maya burials or artistic representations (see Houston 2001). A man, seated against the wall of the grave, was accompanied by a woman at his right side, with her left arm wrapped around his shoulders. In the bend of her right knee lay a newborn. Because their position in death evoked obvious sentiment, they became known as "The Loving Couple."

There is universal curiosity about how closely ancient people resembled us in their feelings and experience of the world. Bioarchaeological research has been quite effective in reconstructing similarities and differences in the physical characteristics of people (Buikstra and Beck 2006) and in their health, material living conditions, and technological abilities, but understanding the personal lives of ancient people presents a greater challenge. Our ability to identify with the sentiment expressed in this couple creates a sense of kinship that compels us to learn more about them. The purpose of this chapter is (1) to reconstruct the biological and social identities of the Lamanai lovers within their cultural context and (2) to promote the movement from traditional osteobiography to social biography for the purpose of better understanding social identity.

Figure 7.1. The burial of N11-5/7A (man in foreground), N11-5/7B (woman with arm around man), and N11-5/7C (baby at woman's knee) at Lamanai, Belize (photo by David M. Pendergast).

Identity and Embodiment in Bioarchaeology

Contemporary postprocessual archaeological theory for some time has fostered the exploration of personal meaning through concepts of agency, identity, and personhood (e.g., Dobres and Robb 2000; Fowler 2004; Hodder 1986; Tilley 1999), but it is still unusual to apply such approaches in osteological analyses. Forms of art and the artifactual traces of people, rather than their actual bodies, have constituted the main sources of data for postprocessual research. Within this paradigm, individuals are viewed as actors who can and do exercise agency to deal with a variety of social contexts, rather than as passive receptors; in addition, the human body is seen as a place for playing out social and political negotiations (Fowler 2004; Joyce

2005; Thomas 1996; Yates 1993). Evidence for these interactions is found in body disposal facilities, grave goods, and body treatment. Although cultural behavior has a profound impact on biology (Goodman and Leatherman 1998) and can be read in skeletons (e.g., Buikstra and Beck 2006; Larsen 1997), the majority of postprocessual burial analyses have been informed mainly by the mortuary body, as an object that has been culturally treated, rather than the biological body. In this chapter we examine the way in which identity and personhood are constituted in current archaeological theory and how they are investigated. We advocate the integration of biological life history with mortuary data for enhancing reconstructions of social identity.

Chris Fowler (2004) states that personhood, or identity, is realized through relationships with things, places, nature, and other humans. Although both biological and social aspects of identity are observable in the body, the use of biological identity in postprocessual archaeology does not often extend beyond the social meaning of age and sex in particular contexts, such as activity restrictions placed on age groups or a particular gender. Although it is understood by bioarchaeologists that culture plays a role in creating the forensic individuation of osteobiography, the social meaning of personal identities is often forgotten in favor of epidemiological meaning. We argue that osteobiographical data such as individual histories of food consumption, disease experience, physical activity, and movement across the landscape should be put to greater use for inferring social identity.

Social identity is derived from social and political interactions and plays a major role in the formation of personal biography and identity (Fowler 2004). The body is typically used to reconstruct identity through analyses of *in vivo* modifications of its (1) surface (e.g., tattoos, piercing, and scarification), (2) form (e.g., parts such as foot binding, cranial modification, or the entire body, such as weight gain or loss), or (3) composition (e.g., highly selective diets) (see also Duncan, this volume, and Torres-Rouff, this volume). Interpretations of identity are typically derived from preserved text and pictorial media (Joyce 2000; Meskell 1999; Meskell and Joyce 2003) and from ethnographic analogy. The bodily incorporation of social meaning and personal expression through these kinds of symbolic embellishment and alteration is referred to as "embodiment" (Csordas 1990, 1999).

Embodiment can be a conscious or subconscious process, but both result in the use of the body to demonstrate cultural and/or individual identity. Such identities are often meant to be visible and are easily observed in public. Others are hidden from public view and therefore have only personal meaning. Because identities are contextual, they can also change from one point in life to another. Modification of the body is a biocultural act, but social meaning is not simply conferred through surface alteration of tissues or morphological alteration of specific body parts. Embodiment can also mean the incorporation of culture into the deeper structure and composition of the skeleton. Connecting the cognitive and cultural being with the biological one can be accomplished with methodologies that move from gross morphology to increasingly fine levels of analysis, such as microscopic structure and chemical composition.

Fowler (2004) notes that substances used in the processes of cultural embodiment provide expressions of social relations and worldview and may be controlled or distributed differentially. Food is among these substances. It is imbued with ideological and social meaning related to its preparation and presentation. The biological act of eating is therefore also a social act, and the chemical composition of food becomes incorporated in our tissues along with all of its social meaning. Hence we are what we eat both biologically and socially. The food that we eat can be either a conscious or an unconscious expression of identity, and part of that identity is how it positions us in terms of nature, place, and other humans. The most direct and detailed reconstructions of food consumption are done with chemical analyses of tissues (especially isotopic analyses), because dietary inferences made from gross morphology or various imaging techniques depend on malnutrition to manifest abnormality.

Social and biological identities also include individual relationships with place. Dietary regimes are closely tied to the physical environment, but so too is the isotopic composition of the water we drink. The climatic and physiographic variability of different environments results in different oxygen isotopic compositions of water that become incorporated into body water and then into the skeleton. The embodiment of place through local water consumption cannot be considered a conscious or a social act of identity expression, but movement between or among landscapes is almost always a conscious and social act that does become embodied in our skeletal chemistry. Oxygen isotope analysis enables the production of landscape bioarchaeology, which includes the reconstruction of individual histories of geographic relocation, which not only create identity but are almost always tied to some form of social process.

Personal identity is also constituted by what people do and how they physically experience the world, such as their state of health and degree and type of physical activity, both of which can be embodied in skeletons.

Although part of *in vivo* identity can involve the individual perception of health or illness, and state of health is certainly related to what people do, people do not as a rule attempt to embody illness as a biocultural act. Similarly, neither skeletal embodiment of one's occupation through repetitive physical activity nor embodiment of single traumatic events resulting from violence or an accident is normally a conscious act. By contrast, altering the physical appearance of the body for cosmetic or identity reasons is a purposeful biocultural act.

Bodies are found in a wide variety of burial or disposal contexts, often containing items of material culture that can have social biographies of their own. The articulation of personhood with material culture in mortuary contexts can be quite tenuous. For example, although objects may be placed in graves to symbolize individual origins or associations with other locations, such evidence of place must still be indirect. Only the skeleton itself can provide direct evidence of either origin in a different location or geographic relocation during life. Furthermore, one cannot assume that the cultural components of a burial (such as the artifacts, facility, burial position, and location) reflect the status of the deceased in life, for it is the living who dispose of the dead and who decide on the mortuary treatment, including the objects to be included in the burial facility. As Fowler (2004) notes, the "person" is socially dissolved by death and then reconstituted by others.

Methodology: An Introduction

Age and sex determinations, differential diagnoses of pathological conditions, and identification of anomalous muscle use were made using well-established criteria (Buikstra and Ubelaker 1994; Ortner 2003; Stone and Stone 1990). The use of oxygen isotope ratios measured in skeletons to identify diet and geographic relocations is based on the premise that "we are what we eat and drink," because our tissues reflect the isotopic composition of the food we consume and the water we imbibe. Previous publications provide details on the theoretical basis and methodology used in isotopic analyses (White et al. 2000, 2002).

Oxygen Isotope Analysis

Briefly, the oxygen isotope composition of water varies with climate and environment and is incorporated into bones and teeth during the process of mineralization (Longinelli 1984; Luz et al. 1984). Bone continually remodels and reflects the last several years of life; but remodeling is faster in

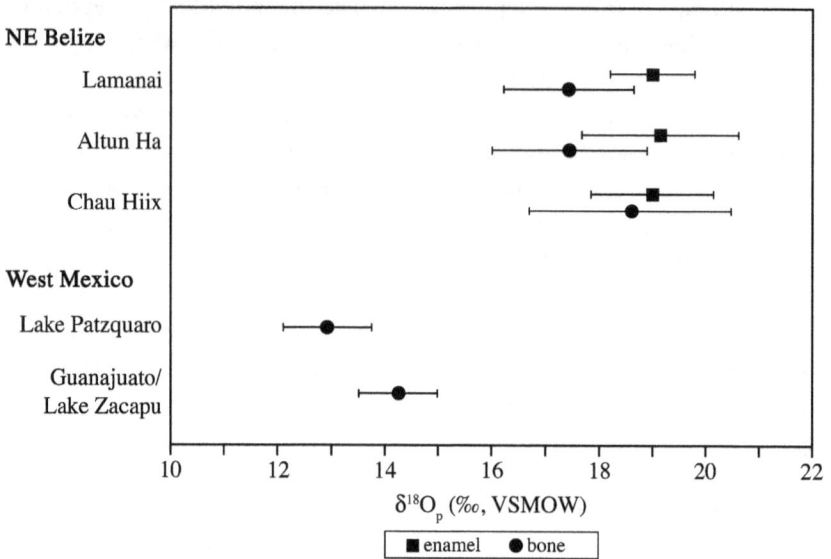

Figure 7.2. Baseline means and ranges of $\delta^{18}O_p$ values for sites in West Mexico and Northeast Belize (VSMOW = Vienna Standard Mean Ocean Water). Lake Patzquaro sites include Tzintzuntzan, Urichu, Tocuaro, Atoyac, and Teremondo. Guanajuato and Lake Zacapu sites include Portales, Guadalupe, Los Nogales, Palacio, and Milpillas.

children, so they will isotopically reequilibrate to new places more quickly than adults. Enamel permanently records water imbibed during its formation, so individual relocations can be identified by comparing enamel with bone or one tooth with another.

Intrasite variability in the Maya area is about 2‰ (per mil) among control samples (figure 7.2) and can be caused by seasonal climate, use of different local water sources, consumption of imported foods with high water content, and the presence of breastfeeding children, who are enriched in ^{18}O because their water source is mother's milk (Wright and Schwarcz 1998; White et al. 2000). Because environments can have overlapping $\delta^{18}O_p$ values, this single line of evidence is not always sufficient for a definitive identification of regions of origin.

Phosphate ($\delta^{18}O_p$) is analyzed here because it is generally better preserved than carbonate (McArthur and Herczeg 1990). Based on the mean crystallinity index (CI) for bone (2.8 ± 0.11) and enamel (3.0 ± 0.13), and the lack of correlation between $\delta^{18}O_p$ values and CI (Pearson's r_{bone} = 0.366, n = 8; Pearson's r_{enamel} = 0.385, n = 7), we infer that recrystallization did not adversely affect the $\delta^{18}O_p$ values (table 7.1). A lack of correlation between $\delta^{18}O_p$ values

Table 7.1. Isotopic Data for "The Loving Couple" and Some Postclassic Contemporaries at Lamanai, Belize

Burial Number	Sex	Age (yrs)	BONE				ENAMEL						
			$\delta^{18}O$ (‰)	YIELD[a] CO_2	YIELD[b] Ag_3PO_4	CI[c]	TOOTH	$\delta^{18}O$ (‰)	YIELD[a] CO_2	YIELD[b] Ag_3PO_4	CI[c]	$\delta^{13}C$ (‰)	$\delta^{15}N$ (‰)
Contemporary Burials													
N10-1/2	M	50–60	17.4		1.5	2.8	M1	19.2	5.0	1.7	3.1	-9.8	10.0
N10-2/40	F	A	16.4	4.9	1.4	2.8	M2	19.2	4.9	1.6	2.9	-9.0	10.1
N10-2/42	F	A	16.3	5.0	1.4	2.7	M2	19.8	4.8	1.6	3.1	-9.1	8.9
N10-4/31	F	A	17.5	4.9	1.4	2.7	M1	18.3	4.8	1.7	2.9	-9.2	9.2
N10-4/31	F	18+	18.6	4.9	1.4	2.9	M3	19.6	4.7	1.8	3.2	-9.9	8.5
Mean			17.2		1.4	2.8		19.2				-9.4	9.3
St. Dev.			0.9					0.6				0.4	0.7
Loving Couple													
N11-5/7A	M	35–50	17.4	4.9	1.4	2.8	M3	19.5	4.7	1.8	2.9	-10.1	9.6
N11-5/7B	F	20–35	16.9	5.0	1.2	3.0	M3	18.7	4.8	1.7	3.1	-9.0	9.2
N11-5/7C	U	NB	16.5	5.0	1.2	2.7	NA	NA	NA	NA	NA	NA	NA

Source: $\delta^{13}C$ and $\delta^{15}N$ values from White and Schwarcz (1989).

[a] CO_2 YIELD (in μ moles/mg Ag_3PO_4).

[b] Ag_3PO_4 = silver phosphate. Ag_3PO_4 YIELD (in mg produced/mg starting material).

[c] CI = crystallinity index.

and phosphate yield (Pearson's r_{bone} = 0.346, n = 8; Pearson's r_{enamel} = 0.038, n = 7) also indicates the absence of preferential recovery of one isotope over the other during phosphate precipitation.

Although diagenesis does not appear to have altered the oxygen isotope composition of this material, the $\delta^{18}O_p$ values of enamel are consistently higher than those of bone by 1–2‰. A bone-enamel difference would normally be interpreted as evidence for relocation, but it is improbable that everyone in this sample relocated. Because these data create some uncertainty about sample integrity, we interpret the $\delta^{18}O_p$ values with caution, particularly those from the more diagenetically susceptible bone.

Carbon and Nitrogen Isotope Analysis

Reconstruction of ancient diets begins with isotopic variation that exists among plants and trophic levels (for a more detailed review of the principles and interpretation of paleodietary isotopic analysis, see Ambrose 1993). Since most humans are omnivores, their collagen $\delta^{13}C$ values ($\delta^{13}C_{col}$) reflect the plants and animals they consume. All plants fall into one of three photosynthetic categories (C_3, C_4, CAM) that have different $\delta^{13}C$ values. Most wild plants, trees, nuts, fruits, and vegetable cultigens are C_3 and have the most negative values (modern average is -26.5‰; O'Leary 1988). On the other hand, C_4 plants have less negative $\delta^{13}C$ values (modern average is -12.5‰; O'Leary 1988). Maize is the most regularly consumed C_4 plant in Mesoamerica.

Marine/reef resources have $\delta^{13}C_{col}$ values that emulate C_4 plants. Therefore nitrogen-isotope ratios in collagen ($\delta^{15}N_{col}$) are used to correct possible misinterpretations of diet by establishing the trophic level and source of dietary protein (DeNiro and Epstein 1981; Schoeninger 1985). The lowest $\delta^{15}N$ values are found in legumes and blue-green algae and the highest in marine mammals (Schoeninger 1985).

Although the Lamanai diet was less dependent on maize than were other regional diets (figure 7.3), it consisted of large quantities of C_4 sources that likely also included C_4-consuming animals such as dogs or deer and C_4-like marine resources (Coyston et al. 1999; White and Schwarcz 1989). The $\delta^{15}N_{col}$ values indicate that significant quantities of marine foods were consumed. Given the available diversity of resources, diets were quite uniform (table 7.1).

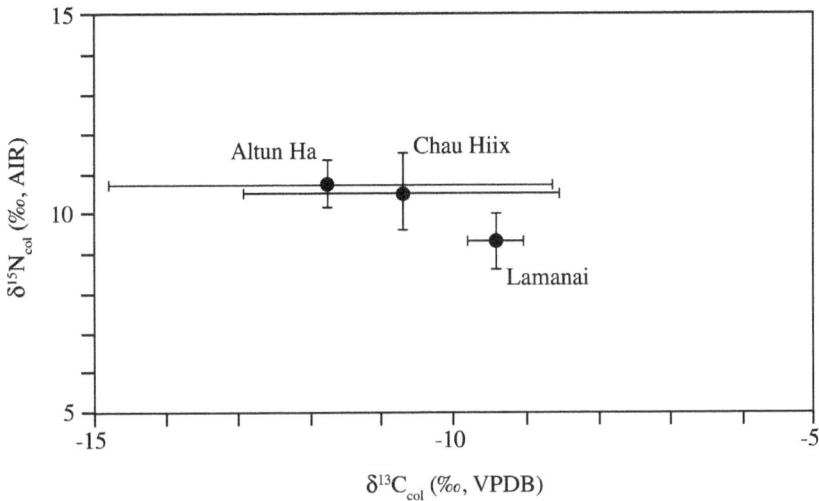

Figure 7.3. Comparison of Postclassic period $\delta^{13}C_{col}$ values for sites in Northeast Belize (VPDB = Vienna Pee Dee Belemnite) (Chau Hiix data from Metcalfe 2005; Altun Ha data from White et al. 2001 and Olsen 2006).

Description of the Skeletons

The Man (N11-5/7A) of "The Loving Couple"

Originally suspended around the man's neck (but fallen onto his lower body) was a pair of copper tweezers of west Mexican style, with circular blades that probably served as a badge of rank or status. They were still displayed this way in western Mexico at the time of Spanish contact. He was also wearing a shell horse-collar ornament on his right forearm (perhaps originally worn on his upper arm). Such ornaments have been found in different contexts elsewhere, but this burial makes clear the way in which they were used (Pendergast 1989).

This man was between 40 and 50 years old and was 164.3 to 166.7 cm tall. An appliance had been fitted to flatten the back of his skull, producing a very broad (brachiocephalic) skull and an extremely vertical occipital region. During excavation, it appeared that his face was deformed (Pendergast 1989), but this impression was not confirmed by subsequent lab analysis. He had, however, suffered from nonspecific maxillary sinusitis, which was in the process of healing at death. Differential diagnosis of healed or healing porotic hyperostosis present on his frontal, parietal, occipital, and zygomatic bones indicates that he may have been iron deficient. Active degenerative joint disease was evident on the articular tubercles of his temporal bone, and there was a mature but benign fibro-osseous tumor in his left mandible between the lower premolar and first molar.

His teeth were in remarkably good condition for his age, although periodontal disease and attrition were pervasive. He had only two carious teeth, no abscesses, and small-to-moderate amounts of calculus. Linear enamel hypoplasia in both upper second incisors records the experience of a systemic stress event when he was approximately 3.5 years old.

His postcranial bones bear witness to trauma and infection that might be associated with his death. The right tibia exhibits nonspecific periosteal reactions related to an ossified subperiosteal hematoma. These were in a state of healing at the time of death and suggest trauma followed by infection. Both fibulae also had nonspecific periosteal reactions resulting from inflammation or infection. There were active lesions in the navicular and medial cuneiform bones of his left hand, for which a differential diagnosis suggests possible mycotic infection.

An examination of his muscle attachment sites indicates that he was engaged in strenuous and repetitive physical activity. He did a lot of work with his upper body and squatted with his feet flat on the ground. His shoulder complex shows that he did something that squeezed his shoulders together. On the medial clavicle there were large attachments of the costoclavicular ligament and large grooves for the subclavius muscle. These are both bilateral but suggestive of greater need for stability, and perhaps trauma, on the right side. On the lateral clavicle, the trapezoid and conoid ligaments were very well developed. Both are involved in absorbing stresses to the upper limb. Attachment sites for the deltoid muscles in the humerus are very well developed in association with stresses on the shoulders, and development of the anterior fibers suggests working with the arms outstretched. All of the muscle attachments on the intertubercular groove of the humerus are very well developed. These muscles are all medial rotators. Of these, the pectoralis major is quite well developed and may have been responsible for keeping the arm in a flexed position while the humerus was rotated. Lack of elbow joint modification supports the premise. Although the upper arm was rotated in either a flexed or outstretched position, lack of elbow joint modification indicates that it was not involved in repetitive bending. In addition to the medial rotation of the humerus, there are massive insertion sites associated with pronation and supination (pronators teres and quadratis) in his forearms (radii, ulnae) (figure 7.4a). There is a bony build-up on these sites (particularly the left) that could be consistent with fracture callus, but radiographs have ruled out this possibility. The presence of some periosteal reaction on his right radius, however, suggests that this region might have been irritated by an external force. Wearing heavy bracelets such as the one he was found with and those seen in artistic depictions of high-status individuals throughout ancient Mesoamerica might have caused this. Lack of radial tuberosity involvement and large supinator crests on the ulna (figure 7.4a) further suggest that his forearm was extended when he supinated his forearms.

In both hands there are huge attachment areas on his first and fifth metacarpals (opponens digiti minimi) for the muscles involved in medio-lateral squeezing of the palm (figure 7.4b). The complementary muscle attachment area on the thumb was not nearly as large. In conjunction with development of the attachment sites for the flexor digitorum superficialis and profundus, the action would be consistent with a grip similar to that used with a screwdriver. This suite of anomalies is consistent with the application of either tensile or compressive forces through relatively straight arms, while gripping something and turning both the upper and lower arms medially.

Muscle markings on the lower limb are a little easier to interpret. He squatted repetitively with the full sole of his foot planted on the ground. The only observable (right) tibia has a well-defined squatting facet on the anterior aspect of the distal metaphyseal area (figure 7.4c). Further evidence for squatting is found in the muscle attachment sites involved in frequent rising from a squatting position. These include bilaterally large insertion sites of the gluteus maximus, well-defined gluteal tuberosities, tibial tuberosities associated with the quadriceps femoris, bilateral enthesiopathies on the anterior surface of the patella, and hyperdevelopment of the soleus muscle attachment areas on the fibula.

There is also an interesting example of activity-related dental wear between his upper and lower first left incisors. This location shows evidence of localized wear consistent with continual abrasion from a hard, rounded substance, such as cording, as it is passed between the teeth while they are closed. This anomaly strongly suggests that he used his dentition as a tool. Combined with the backward stretch of the head and neck evident in the clavicle, it appears that he may have been gripping and pulling some kind of cord with his incisors.

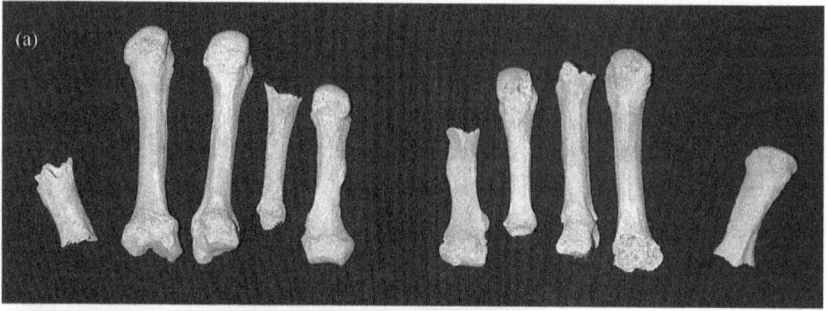

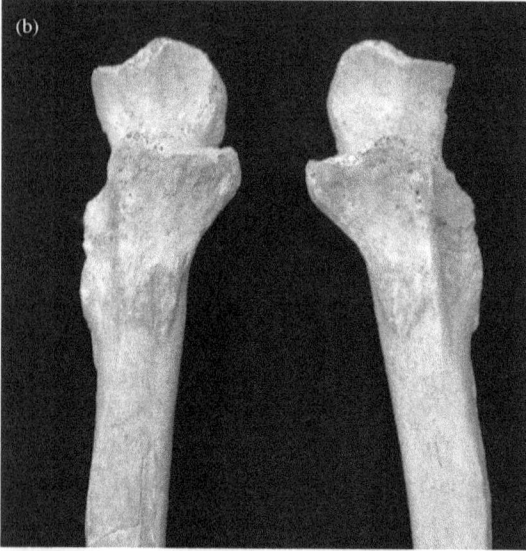

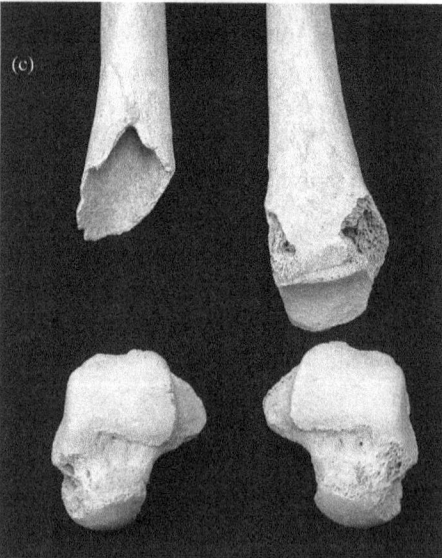

Figure 7.4. Muscle markings in (a) metacarpals, (b) ulnae, and (c) squatting facets (photos by Jay Maxwell).

The Woman (N11-5/7B) of "The Loving Couple"

Like her companion in death, the woman also shows cultural evidence of west Mexican origin. Her hair had been formed into a braid or queue, bound in cloth (since largely decayed), and encased in a group of five copper-tin bronze rings. This hairstyle was common in ancient western Mexico and is still seen there today (Dorothy Hosler, personal communication, 1989).

The woman was much younger, between 20 and 25 years old, and 152.1 to 153.9 cm tall. Although her skull is fragmentary, the form of the remaining elements does suggest the same style of cultural modification exhibited by her companion. Poor preservation precludes any further observation of pathology in the cranial bones, but 30 of her teeth were preserved. They show minimal wear, periodontal disease, and calculus. None of her teeth have abscesses, and only 3 have caries. The location of linear enamel hypoplasia in all of her second and third molars indicates that she experienced a severe stress event when she was just over six years of age and another around the age of ten.

Her right fibula shows nonspecific periosteal reaction that was healing at the time of death and is suggestive of some form of infection. The interfaces between lumbar vertebrae 1 and 2 and lumbar vertebrae 3 and 4 have lesions that could represent either herniations or Schmorl's nodes, but it is unclear if they were active or healed at death. If healed, they are more likely to be Schmorl's nodes and represent some form of exertion during adolescence, such as carrying heavy loads on the back. In addition, an active resorptive lesion in the eleventh thoracic vertebra may indicate the early stages of infection or cancer.

The Baby (N11-5/7C) of "The Loving Couple"

Bone measurement indicates that the age of this infant was about 34 weeks of gestation, but the child shows signs of a systemic infection that could have resulted in a growth deficit. Both orbits exhibit multiple layers of new woven bone that also covers 33–75% of the surfaces of all observable long bones and the left ilium. It is likely that the nonspecific infection in the mother's skeleton was transferred to the baby.

Discussion

The placement and age of the baby would suggest that it had been born. There are two other lines of evidence indicating that the young woman was the mother. Their bone oxygen isotopic compositions are very similar (table

7.1—the small difference could be accounted for by analytic error), and they both show signs of infection. The healing fibula and the possible infection in the woman's vertebra, combined with the extent of infection found in the baby, indicate that she had been fighting illness for a long time. At 34 weeks, the baby would have been premature but viable. It is not possible to determine if it was stillborn or died shortly after birth. Its sickened state, however, could have resulted in lower than normal weight and height, making it older than 34 weeks, possibly even full term. The fact that the baby was not a secondary burial suggests that it died around the same time as the mother. Furthermore, being buried with its mother in this special context indicates that despite its newborn state it (and perhaps babies in general) had status and value in the eyes of the community and that the mother-infant bond was recognized as important. This baby did not live long enough to establish or embody an identity on its own or act as an agent, but its presence and position in this burial indicate that even as a baby it had a social identity.

There are many conceivable circumstances for the cause of the mother's death. The most natural speculations, however, involve complications from childbirth or the observed infection(s). Other explanations for her demise cannot be ruled out, but if she met a violent death such as suicide, execution, or murder it did not leave marks.

In terms of a biosocial persona, she was probably not high in status, and her hairstyle indicates that she may have come from western Mexico. The two acute health crises in her childhood may have made her "frail" or more susceptible to disease and early death, as per the predictions of the Osteological Paradox of James Wood et al. (1992). Although it is unclear whether some of the lesions in her lumbar vertebrae are a consequence of infection or of carrying heavy loads during her adolescence, either could have added to her frailty. The possibility that she regularly bore heavy loads on her back evokes a more specific persona for her social identity, one that is also consistent with her poor childhood health and suggests origins from a relatively low social status. Furthermore, her stature at 152.1 to 153.9 cm is lower than female averages for the region and time period, which range from 155.1 to 156.4 cm (Glassman and Garber 1999; O'Neal 1997). And yet in death she was placed in a location that would not normally be attributed to a low-status individual. It is possible that she had gained status from foreign identity or from her association with the man around whom her arm was draped. None of her isotopic values are anomalous with individuals assumed to be local to Lamanai. Previous research (Coyston et al. 1999; White and Schwarcz 1989) has demonstrated that the Lamanai diet has distinctively high $\delta^{15}N$ values, reflecting its proximity to aquatic and marine resources. Although west Mexican diets were also maize based, they did not contain the marine component found at Lamanai (Cahue 1999). Similarly, the oxygen-isotope compositions of both her bone and third molar do not match those of any west Mexicans yet analyzed (unpublished data).

There is an intriguing conflict between the chemical composition (δ values) of this woman's tissues and the forms of cultural embodiment (such as hairstyle and possibly cranial modification) that she expresses. We can only speculate on the reasons. The most likely explanation is that she was brought to the site at a very early age, before nine years, when her third molar began to form. She might also have been a second-generation (or more) immigrant, maintaining her identity for honor or simply copying a style she admired (although we assume that the latter behavior may not have been generally tolerated).

The social identity of the man is just as enigmatic. The skeletal morphology and material culture evidence would suggest that he was a foreign elite individual who engaged in a specific repetitive physical activity. He had survived trauma to his lower limbs long enough for infection to appear.

The active maxillary infection suggests that he was not in very good health at the time of his death, and the presence of iron deficiency indicates a past, and possibly ongoing, chronic condition. Dietary insufficiency is only one of many causes of iron deficiency anemia. Among the Maya, this condition is often associated with heavy maize dependency (for a review, see Wright and White 1996), but his $\delta^{13}C$ value indicates that he consumed the least amount of maize among his comparison group, none of whom exhibited the same lesions. Insufficient iron consumption is also unlikely, because his $\delta^{15}N$ value indicates that he would have consumed considerable meat (including marine resources) from a fairly high trophic level. In fact, he had the highest $\delta^{15}N$ value in this sample. It is possible that his iron deficiency was caused by parasitic infection associated with the consumption of aquatic and marine resources and common in this area in both ancient and modern times (White et al. 2006). The status indicated by his copper adornment did not buffer him from this condition and may, in fact, have exacerbated it. Nonetheless, the anemia does not appear to have seriously affected his growth and development, as he was just slightly taller (at 164.3 to 166.7 cm) than his male contemporaries of the region (163.9 to 165.2 cm) (see Danforth 1994; O'Neal 1997).

Cranial modifications are an expression of group rather than individual identity, as they are an adult intervention in childhood growth (see also

Duncan, this volume; Torres Rouff, this volume). The lambdoidal flattening cranial modification stands out against the dominant fronto-occipital style practiced at Lamanai (White 1996), which suggests that he may in fact have been born elsewhere. This style of cranial modification has been reported for sites in Michoacán (Carot 2001; Pereira 1999), so both the cranial modification and the west Mexican style of tweezers that he wore point to a west Mexican origin. His seemingly normal-to-tall body height in the face of nutrient deficiency could also be explained by origins from a genetically taller population. Unlike his partner, he shows no other sign of childhood stress except for a hypoplastic dental defect consistent with weaning stress around the age of 3.5 years. The inability of status to buffer weaning stress is also found at other Maya sites (see Storey 1992). As one of the oldest individuals in the sample, it is remarkable that he lived as long as he did (40–50 years).

As with the woman's, the man's diet and water consumption are consistent with the local population. If he did come from elsewhere, he must have arrived before his third molar began to develop (i.e., under nine years of age, like his mate). His possession of foreign high-status adornment and his young arrival at Lamanai would suggest that his status was ascribed rather than achieved, assuming that his identity was not redefined by those who buried him. Although his physical characteristics could not have been altered after death, they are equivocal. The cranial modification indicates residence in a foreign location at a very early age. The diet indicates possible continued high status after arrival. His iron deficiency could be consistent with a high-status diet, but his infection is probably not.

His survival to "old" age is consistent with high status, but his strenuous physical activity is unexpected. Although his muscle markings indicate occupational activity, we cannot be sure that the actions involved were carried out simultaneously. Nonetheless, one activity that could explain his suite of activities is weaving. Although women were primarily responsible for weaving fabric in ancient times (e.g., Beaudry-Corbett and McCafferty 2004; McCafferty and McCafferty 1998) as they are today, it is possible that men were involved in weaving elite products, such as the heavier mats that symbolized kingship and were used on thrones. Heavy fibers would have been necessary for mat weaving, which could account for the strong muscle markings. These could have included imported maguey, as used in Mexico (Parsons and Parsons 1990), or perhaps local palm. Linda Manzanilla (personal communication, 2006) has postulated that male weavers in Mexico were engaged in the production of high-status or specialized products, but there is also some evidence of elite men being involved in such a craft. For example, a male weaver, identified by a weaving needle, was found among foreign individuals at the Merchants' Barrio, Teotihuacan (Spence et al. 2005). This activity would also be consistent with Takeshi Inomata's (2001) recent suggestion that Maya elites produced specialized crafts that functioned to concentrate their prestige, political power, and artistic aesthetic and further separate them from lower-status individuals who were responsible for the majority of craft production. He suggests that elite products were imbued with ideology, power, and esoteric knowledge and used as currency for elite competition. Inomata views them as a hyper-expression of group identity that illustrates Pierre Bourdieu's (1984) concept of cultural capital for enhancement.

Although much attention is given to artifacts in mortuary contexts, little has been paid to using the skeleton to identify their possible producers. The elite-crafted products described by Inomata (2001) were made of substances that preserve well, such as shell and lithics. The importance of organic products is not reflected in the archaeological record because of poor preservation, but the body of N11-5/7A illustrates the potential of the skeleton to bear witness to the creation of a product that has not survived. Even though N11-5/7A had high status, he may have been weaving or making mats for someone of higher rank, and his product(s) could have been buried with the receiver. Hence only his body would remain to signal his identity.

Conclusions

The investigation of this "family" raises as many questions about their identities as it does answers. They form a meaningful social unit with a distinct identity among the other burials at the site, but each individual has its own biosocial identity. We still cannot explain why they were given such unusual mortuary treatment (including location and particularly body position), whether it reflected their position in life, whether the expression of sentiment imposed on their bodies after death was a reflection of their feelings during life, exactly what kind of activity the man was engaged in so repetitively, why they were living at Lamanai, and what caused them to die at the same time and be buried together. The combination of biological and material culture data does, however, allow us to make some fairly safe assumptions.

All suffered some ill health at the time of death, which is likely to have been the cause of death of the baby. Both the man and the woman were probably foreigners who were brought from west Mexico at an early age. The man had high status when he arrived and was able to maintain it even in a foreign land, perhaps because he was a highly specialized artisan with esoteric knowledge. The age difference between him and the woman indicates that he must have been in the community for about 20 years before she arrived, which hints at the possibility of a long-term relationship between the two regions. One wonders if she was brought from their homeland at his request.

Although the couple became dietarily assimilated into their new biocultural environment, their artifacts indicate that they preserved some symbols and practices of their homeland culture. The man was regularly engaged in a repetitive physical activity that probably had something to do with his social or political role in the community. Alterations of his skeleton indicate that he may have been a weaver of heavy fibers. Both the man and the woman had also experienced some trauma to the lower limbs, which they had survived for some time before death. The limitations of isotopic data as a basis for identifying foreigners are also illustrated by this case. Even in the absence of artifactual evidence, isotopic compositions have been used successfully in many Mesoamerican cases to demonstrate variable geographic origins and life-history movements across landscapes (Spence et al. 2004, 2005; White et al. 1998, 2000, 2002; White, Longstaffe, and Law 2001; White, Pendergast, et al. 2001; White, Spence, et al. 2004; White, Storey, et al. 2004). Here, however, they are only able to provide evidence of long-term residence at the site.

These burials have features (including artifacts and cranial modification) that represent conscious acts of cultural identity and embodiment, but the physical activities, health, and chemical composition of the individuals constitute unconscious expressions of embodiment that are no less revealing of culture. Indeed, the very fact that they are not intentional implies a sort of objectivity that can only be left to science.

Acknowledgments

We thank Jaime Awe and the Department of Archaeology, Belize, for the opportunity to examine these skeletons, Michael Spence and Janet Gardner for their helpful commentary on the muscle markings of the skeletons, Barbara Hewitt for the graphs, and Rethy Chemm for assisting with interpretation of the x-rays. This is the Laboratory for Stable Isotope Science publication 233.

References Cited

Ambrose, Stanley H. 1993. "Isotopic Analysis of Paleodiets: Methodological and Interpretive Considerations." In *Investigations of Ancient Human Tissue: Chemical Analyses in Anthropology*, edited by M. K. Sandford, 59–130. Langhorn, Penn.: Gordon and Breach.

Beaudry-Corbett, Marilyn, and Sharisse McCafferty. 2004. "Spindle-Whorls: Household Specialization at Ceren." In *Ancient Maya Women*, edited by T. Ardren, 52–67. Walnut Creek, Calif.: AltaMira Press.

Bourdieu, Pierre. 1984. *Distinction: A Social Critique of the Judgement of Taste*. Translated by R. Nice. Cambridge, Mass.: Harvard University Press.

Buikstra, Jane E., and Lane A. Beck, editors. 2006. *Bioarchaeology: The Contextual Analysis of Human Remains*. Amsterdam: Elsevier.

Buikstra, Jane E., and Douglas H. Ubelaker, editors. 1994. *Standards for Data Collection from Human Skeletal Remains*. Arkansas Archeological Survey Research Series No. 44. Fayetteville: Arkansas Archeological Survey.

Cahue, Laura. 1999. "Environmental Change, Sociopolitical Power and the Diet and Health of Elites during the Emergence of the Tarascan State." *American Journal of Physical Anthropology Supplement* 28: 101.

Carot, Patricia. 2001. *Le Site de Loma Alta, Lac de Zacapu, Michoacán, Mexique*. Paris Monographs in American Archaeology 9. Series editor, Eric Taladoire. BAR International Series 920. Oxford: Archaeopress.

Coyston, Shannon, Christine D. White, and Henry P. Schwarcz. 1999. "Dietary Carbonate Analysis of Bone and Enamel for Two Sites in Belize." In *Reconstructing Ancient Maya Diet*, edited by C. D. White, 199–220. Salt Lake City: University of Utah Press.

Csordas, Thomas. 1990. "Embodiment as a Paradigm for Anthropology." *Ethos* 18: 5–47.

———. 1999. "Embodiment and Cultural Phenomenology." In *Perspectives on Embodiment: The Intersections of Nature and Culture*, edited by G. Weiss and H. Faber, 143–162. London: Routledge.

Danforth, Marie. 1994. "Stature Change in Prehistoric Maya of the Southern Lowlands." *Latin American Antiquity* 5: 206–211.

DeNiro, Michael J., and Samuel Epstein. 1981. "Influence of Diet on the Distribution of Nitrogen Isotopes in Animals." *Geochimica et Cosmochimica Acta* 45: 341–351.

Dobres, Marcia-Anne, and John Robb. 2000. *Agency in Archaeology*. New York: Routledge.

Fowler, Chris. 2004. *The Archaeology of Personhood: An Anthropological Approach*. London: Routledge.

Glassman, David M., and James F. Garber. 1999. "Land Use, Diet, and Their Effects on the Biology of the Prehistoric Maya of Northern Ambergris Cay, Belize." In *Reconstructing Ancient Maya Diet*, edited by C. D. White, 119–132. Salt Lake City: University of Utah Press.

Goodman, Alan H., and Thomas L. Leatherman, editors. 1998. *Building a New Biocultural Synthesis: Political-Economic Perspectives on Human Biology*. Ann Arbor: University of Michigan Press.

Hodder, Ian. 1986. *Reading the Past: Current Approaches to Interpretation in Archaeology.* Cambridge: Cambridge University Press.

Houston, Stephen D. 2001. "Decorous Bodies and Disordered Passions: Representations of Emotion among the Classic Maya." *World Archaeology* 33: 206–219.

Inomata, Takeshi. 2001. "The Power and Ideology of Artistic Creation." *Current Anthropology* 42: 321–349.

Joyce, Rosemary. 2000. *Gender and Power in Prehispanic Mesoamerica.* Austin: University of Texas Press.

———. 2005. "Archaeology of the Body." *Annual Review of Anthropology* 34: 139–158.

Larsen, Clarke S. 1997. *Bioarchaeology: Interpreting Behavior from the Human Skeleton.* Cambridge: Cambridge University Press.

Longinelli, Antonio. 1984. "Oxygen Isotopes in Mammal Bone Phosphate: A New Tool for Paleohydrological and Paleoclimatological Research?" *Geochimica et Cosmochimica Acta* 48: 385–390.

Luz, Boaz, Yehoshua Kolodny, and Michal Horowitz. 1984. "Fractionation of Oxygen Isotopes between Mammalian Bone-Phosphate and Environmental Drinking Water." *Geochimica et Cosmochimica Acta* 48: 1689–1693.

McArthur, J. M., and A. Herczeg. 1990. "Diagenetic Stability of the Isotopic Composition of Phosphate-Oxygen: Paleoenvironmental Implications: Phosphorite Research and Development." *Geological Society Special Publication* 52: 119–124.

McCafferty, Sharisse, and Geoffrey G. McCafferty. 1998. "Spinning and Weaving as Female Identity in Post-Classic Mexico." In *Reader in Gender Archaeology*, edited by K. Hays-Gilpin and D. S. Whitley, 213–230. New York: Routledge.

Meskell, Lynn. 1999. *Archaeologies of Social Life: Age, Sex, Class Et Cetera in Ancient Egypt.* Social Archaeology Series. Oxford: Blackwell.

Meskell, Lynn, and Rosemary Joyce. 2003. *Embodied Lives: Figuring Ancient Maya and Egyptian Experience.* London: Routledge.

Metcalfe, Jessica. 2005. "Diet and Migration at Chau Hiix, Belize: A Study of Stable Carbon, Nitrogen, and Oxygen Isotopes." Master's thesis, University of Western Ontario, Department of Anthropology.

O'Leary, Marion. 1988. "Carbon Isotopes in Photosynthesis." *Bioscience* 38: 328–336.

Olsen, Karyn. 2006. "Dedication and Sacrifice: An Oxygen-Isotope Study of Human Remains from Altun Ha, Belize, and Iximché, Guatemala." Master's thesis, University of Western Ontario, Department of Anthropology.

O'Neal, Paul J. 1997. "Skeletal Measurements and Stature Comparisons in Two Maya Populations from Belize." Honors thesis, University of Western Ontario, Department of Anthropology.

Ortner, Donald. 2003. *Identification of Pathological Conditions in Human Skeletal Remains.* 2nd ed. Amsterdam: Academic Press.

Parsons, Jeffrey, and Mary H. Parsons. 1990. *Maguey Utilization in Highland Central Mexico: An Archaeological Ethnography.* Anthropological Papers, No. 82. Ann Arbor: University of Michigan, Museum of Anthropology.

Pendergast, David M. 1989. "The Loving Couple: A Mystery from the Maya Past." *Archaeological Newsletter*, series II, 30: 1–4. Toronto: Royal Ontario Museum.

Pereira, Grégory. 1999. *Potrero de Guadalupe: Anthropologie funéraire d'une commu-nauté pré-tarasque du nord du Michoacán, Mexique*. Paris Monographs in American Archaeology 5. Series editor, Eric Taladoire. BAR International Series 816. Oxford: Archaeopress.

Schoeninger, Margaret J. 1985. "Trophic Level Effects on $^{15}N/^{14}N$ and $^{13}C/^{12}C$ Ratios in Bone Collagen and Strontium Levels in Bone Mineral." *Journal of Human Evolution* 14: 515–525.

Spence, Michael, Christine D. White, Fred J. Longstaffe, and Kimberley R. Law. 2004. "Victims of the Victims: Human Trophies Worn by Sacrificed Soldiers from the Feathered Serpent Pyramid, Teotihuacan." *Ancient Mesoamerica* 15: 1–15.

Spence, Michael, Christine D. White, Evelyn C. Rattray, and Fred J. Longstaffe. 2005. "Past Lives in Different Places: The Origins and Relationships of Teotihuacan's For-eign Residents." In *Early Civilizations, Settlement and Subsistence: Essays in Honor of Jeffrey R. Parsons*, edited by R. E. Blanton, 155–197. Los Angeles: Cotsen Institute of Archaeology, University of California, Los Angeles.

Stone, Robert J., and Judith A. Stone. 1990. *Atlas of the Skeletal Muscles*. Dubuque, Iowa: Wm. C. Brown.

Storey, Rebecca. 1992. "The Children of Copán: Issues in Palaeopathology and Palaeode-mography." *Ancient Mesoamerica* 3: 161–167.

Thomas, Julian. 1996. *Time, Culture and Identity: An Interpretive Archaeology*. London: Routledge.

Tilley, Christopher. 1999. *Metaphor and Material Culture*. Oxford: Blackwell.

White, Christine D. 1996. "Sutural Effects of Fronto-Occipital Cranial Modification." *American Journal of Physical Anthropology* 100: 397–410.

White, Christine D., Fred J. Longstaffe, and Kimberley R. Law. 2001. "Revisiting the Teotihuacan Connection at Altun Ha: Oxygen-Isotope Analysis of Tomb F-8/1." *Ancient Mesoamerica* 12: 65–72.

White, Christine D., Fred J. Longstaffe, Michael W. Spence, and Kimberley R. Law. 2000. "Testing the Nature of Teotihuacan Imperialism at Kaminaljuyu Using Phosphate Oxygen-Isotope Ratios." *Journal of Anthropological Research* 56: 535–558.

White, Christine D., Jay Maxwell, Alexis Dolphin, Jocelyn Williams, and Fred J. Long-staffe. 2006. "Pathoecology and Paleodiet in Postclassic/Historic Maya from North-ern Coastal Belize." *Memorias do Instituto Oswaldo Cruz* 101: 35–42.

White, Christine D., David M. Pendergast, Fred J. Longstaffe, and Kimberley R. Law. 2001. "Social Complexity and Food Systems at Altun Ha, Belize: The Isotopic Evi-dence." *Latin American Antiquity* 12: 371–393.

White, Christine D., and Henry P. Schwarcz. 1989. "Ancient Maya Diets, Inferred from Isotopic and Elemental Analysis of Human Bone." *Journal of Archaeological Science* 16: 451–474.

White, Christine D., Michael W. Spence, Fred J. Longstaffe, and Kimberley R. Law. 2004. "Demography and Ethnic Continuity in the Tlailotlacan Enclave of Teotihuacan: The Evidence from Stable Oxygen Isotopes." *Journal of Anthropological Archaeology* 23: 385–403.

White, Christine D., Michael W. Spence, Fred J. Longstaffe, Hilary Stuart-Williams, and

Kimberley R. Law. 2002. "Geographic Identities of the Sacrificial Victims from the Feathered Serpent Pyramid, Teotihuacan: Implications for the Nature of State Power." *Latin American Antiquity* 13: 217–236.

White, Christine D., Michael W. Spence, Hilary Le Q. Stuart-Williams, and Henry P. Schwarcz. 1998. "Oxygen Isotopes and the Identification of Geographical Origins: The Valley of Oaxaca versus the Valley of Mexico." *Journal of Archaeological Science* 25: 643–655.

White, Christine D., Rebecca Storey, Fred J. Longstaffe, and Michael W. Spence. 2004. "Immigration, Assimilation, and Status in the Ancient City of Teotihuacan: Stable Isotopic Evidence from Tlajinga 33." *Latin American Antiquity* 15: 176–198.

Wood, James W., George R. Milner, Henry C. Harpending, and Kenneth M. Weiss. 1992. "The Osteological Paradox: Problems of Inferring Prehistoric Health from Skeletal Samples." *Current Anthropology* 33: 343–370.

Wright, Lori E., and Henry P. Schwarcz. 1998. "Stable Carbon and Oxygen Isotopes in Human Tooth Enamel: Identifying Breastfeeding and Weaning in Prehistory." *American Journal of Physical Anthropology* 106: 1–18.

Wright, Lori E., and Christine D. White. 1996. "Human Biology in the Classic Maya Collapse: Evidence from Paleopathology and Paleodiet." *Journal of World Prehistory* 10: 147–198.

Yates, T. 1993. "Frameworks for the Archaeology of the Body." In *Interpretive Archaeology*, edited by C. Tilley, 31–72. Oxford: Berg.

8

Cranial Modification among the Maya

Absence of Evidence or Evidence of Absence?

WILLIAM N. DUNCAN

There is increasing consensus among researchers interested in cranial modification in Mesoamerica and among the Maya in particular that the practice
was a normal part of growing up in Maya society (Duncan and Hofling
2004; Geller 2004; Joyce 2001; Tiesler Blos 1998, 1999). Using ethnohistoric
sources, researchers have argued that the biological body needed to be acted
on in various ways in order to become a fully functioning social entity in
Maya society. This process included head binding as a childhood rite of passage. This is consistent with early documentation suggesting that the practice was widespread (Tozzer 1941). Studies within the Maya area, however,
have found frequencies of modified crania to range from 88% of observable
crania in the largest survey (Tiesler Blos 1999) to slightly over 55% in some
individual sites (Saul and Saul 1997). This raises several questions. Are individuals without modified crania somehow less fully embodied members of
society than individuals with such modification? How might we reconcile
the archaeological and ethnohistoric records?

In this chapter I argue that absence of cranial modification does not constitute evidence of absence of embodiment. First, I briefly review the history of cranial modification studies in Mesoamerica and specifically in the
Maya area, outlining current thoughts about the practice. In particular, I
review ethnographic evidence on childbirth and rites of passages suggesting
that head binding was a normal step in the embodiment process for Maya
children. Second, I review data to show that the frequencies of modified
skulls in the archaeological record indicate that not everyone was the re-
cipient of cranial modification. Finally, I argue that the archaeological and
ethnohistoric records may be reconciled by considering the state of children
when their heads were modified. Specifically, young children were vulnerable to injury from evil winds and susceptible to losing animating essences
out of the tops of their heads (soul loss). Metaphoric polysemic associations between the human body and elements of houses suggest that binding

infants' heads was tantamount to building a roof on top of the children's heads in an attempt to protect them. Binding the head and modifying the shape of the head were often part of the same rite of passage for the Maya, but the two phenomena are not identical. Acknowledging the point allows us to reconcile the archaeological and ethnographic records.

Cranial Modification Studies: An Introduction

Studies of cranial modification have traditionally fallen under one of the following four categories: description and classification of types of modification, assessment of the influence of modification on health, quantification of morphological changes stemming from the practice, and identification of the social causes, correlates, and functions of the cranial modification (after Torres-Rouff 2003). The worldwide history of cranial modification has been reviewed recently by several authors (Duncan and Hofling 2004; Gerszten and Gerszten 1995; Goodrich and Tutino 2001; Tiesler Blos 1998; Torres-Rouff 2003; Tubbs et al. 2006) and does not need to be repeated here. Suffice it to say that documentation of and speculation about the classification, causes, and consequences of the practice stretch back at least as early as the writings of Hippocrates (400 BC [1952]).

Cranial Modification Studies in Mesoamerica

Research investigating cranial modification in Mesoamerica has followed a trajectory similar to that of other regions. The earliest reports of the practice in Mesoamerica were largely descriptive and date to the 1830s, when Edward Mühlenpfort described a case from Oaxaca (Romano Pacheco 1974). Similar descriptions of cranial modification in assemblages throughout Mexico continued into the 1900s (e.g., Dávalos Hurtado 1945, 1946, 1965). Although there have been a number of attempts to classify different types of cranial modification (e.g., Allison et al. 1981; L. A. Gosse, 1855, cited in Dingwall 1931; L. Lunier, 1869, cited in Dingwall 1931; Magitot 1885; Topinard 1879), the vast majority of researchers working in Mesoamerica have adopted the basic classificatory scheme proposed by Adolfo Dembo and José Imbelloni (1938), who classified modifications as either annular (caused by circumferential wrapping) or tabular (due to flattening in one plane). These types were further divided into erect and oblique subtypes. Erect means that the modification is parallel with the neck, while oblique means that the modification is on a plane that would intersect the plane of the neck (see also Torres-Rouff, this volume).

One common theme over time among researchers working in Mesoamerica is the documentation of geographic and temporal trends among different types of cranial modification. This likely reflects the widespread use of the practice in the region and a time-depth stretching back at least 7,000 years (in Tehuacan: Tiesler Blos 1998). Various authors have investigated these questions, but Arturo Romano Pacheco (1974) and Javier Romero Molina (1970) are cited most often with regard to temporal trends. Romano Pacheco (1974) found that tabular erect modifications were the most common in all Pre-Columbian periods in Mexico but that they were most common in the Preclassic (ca. 2500–300 BC) and Postclassic (ca. AD 900–1500) periods. Romero Molina (1970) found that tabular varieties seemed to predate annular varieties. The former date to the Early Preclassic (at some 1400–1200 BC) in the Valley of Mexico, while the latter appear in Oaxaca in the Middle Preclassic (after 1200 BC). Vera Tiesler Blos (1998) found considerable temporal and geographic variability in types of cranial modification, but she notes that the tabular oblique style seems to have been less frequent in the Postclassic. Erect forms were more common overall in the highlands, and oblique forms were more common in the southeastern areas in the lowlands.

Unlike researchers working in other areas (e.g., Allison et al. 1981; Gerszten 1993; see Blackwood and Danby 1955; Flower 1882; and Tommaseo and Drusini 1984 for reviews through time), Mesoamerican researchers have spent less time considering potential health repercussions stemming from cranial modification. While authors in other areas have investigated the general relationship between the practice and health, intelligence, or psychology, researchers in Mesoamerica have circumscribed their attention to more specific questions. In particular, William Feindel (1988: 218) asked if cranial modification could have caused damage to the hippocampus, which he suggested might have caused "memory and other intellectual impairment," leading to the social upheaval around AD 900 that is sometimes known as the Maya collapse. The suggestion was not cited by researchers in a recent volume on the topic (Demarest et al. 2004).

A number of researchers, particularly physical anthropologists, have tried to study the morphological effects of cranial modification (Anton et al. 1992; Bennett 1965; Björk and Björk 1964; Cheverud et al. 1992; Cheverud and Midkiff 1992; Dean 1995; Falkenburger 1938; Frieß and Baylac 2003; Gottlieb

1978; Kohn et al. 1993; Konigsberg et al. 1993; McGibbon 1912; McNeill and Newton 1965; Moss 1958; O'Loughlin 1996, 2004; Ossenberg 1970; Oetteking 1924; Rogers 1975; White 1996). Researchers working in Mesoamerica have also considered the topic. In an effort to quantify the effects of modification, Juan Comas (1960, 1966, 1969; Comas and Marquer 1969; see also Tiesler Blos 1995) studied the effect of the practice craniometrically in a series of studies in the 1960s.

Although all of the previously discussed approaches to studying cranial modification have been useful for understanding one or more facets of the practice, they have not fully explained why the practice actually took place in Mesoamerica (Buikstra 1997). In the past 10 years researchers have been increasingly interested in the roots of the practice and have attempted to approach it from a social perspective. Tiesler Blos (1999), who analyzed more than 1,000 skulls from over 90 sites, stands out among Mesoamerican researchers as the most comprehensive to date. She found that over 88% of the skulls had been modified and came to several conclusions, which summarize the current state of thought on cranial modification in the region. First, no clear correlation exists between presence or absence of modification or type of modification and vertical social status among the Maya. One ethnohistoric source (Torquemada 1613 [1943]) noted that cranial modification was generally associated with nobility in the New World, and there may be some trends at individual sites such as Copán. More recent work has echoed this, suggesting that royal accession involved wrapping the head and attaching a diadem on the forehead of new rulers (Houston et al. 2006). Stephen Houston's suggestion that the goal of cranial modification was to make the head look like an ear of corn, like the Maize God, may reflect this as well (Houston et al. 2006). These suggestions are cogent and are certainly true in some contexts in the Maya area. When the Maya region is viewed as a whole, however, there are no clear trends demonstrating these ideas on a regional basis (Tiesler Blos 1998, 1999). Second, Tiesler Blos (1998, 1999) argued that cranial modification was used to reinforce horizontal social distinctions, as Christina Torres-Rouff (2002, 2003, this volume) found in South America. This is consistent with ethnohistoric sources (e.g., Diego de Landa [Tozzer 1941]), which reported that the Maya could be distinguished from outsiders in central Mexico in part by the presence or absence of cranial modification. Iconography bears this out: Maya individuals are portrayed with modified skulls, while outsiders are shown as lacking a sloped forehead (Duncan and Hofling 2004). Third, Tiesler Blos (1998, 1999) argued (as did Bonavides

Mateos 1992) that cranial modification was part of normal rites of passages for Maya children.

This last idea has been echoed by other researchers throughout Meso-america. Rosemary Joyce (2000: 475) contended that among the Aztec children were "raw materials that needed to be worked into specific forms." Tiesler Blos (1998) and Pamela Geller (2004) have both noted that cranial modification appears to have been associated with fixing gender and related occupational roles among the Maya. Specifically, Geller (2004: 385) argued that "cranial shaping marked . . . ascribed occupational identity." This is consistent with reports of modern *héetz-méek'* ceremonies, in which children have occupational tools placed in their hands at an early stage (see below). A recent comparison of ethnohistoric and ethnographic descriptions of childbirth and socialization among the colonial and modern Yukatek and modern Tzotzil speakers supports the point (Duncan and Hofling 2004).

Ethnohistoric and Ethnographic Perspectives on Cranial Modification

Ethnohistoric descriptions of the practice in Yucatan by Diego de Landa (Tozzer 1941) indicate that head molding occurred within the first week after birth. Only after head molding would the child be taken to a priest to receive a name and determine the child's occupation and destiny. Modern Yukatek and Tzotzil childbirth and socialization have parallel structures.

In the case of modern Yukatek speakers, baptism occurs early, often in the first month of life (Redfield and Villa Rojas 1934; Villa Rojas 1945). For the first three or four months of life babies were carried across their mothers' arms. During this time, children were particularly at risk to be harmed, in part because their souls were not firmly fixed in their bodies (as discussed below). At three or four months, children went through the *héetz-méek'* ceremony. During this ceremony, the godparent of the same sex as the child placed the child astride his or her hip. The godparent then gave the child certain objects associated with the child's future work. Boys were given machetes, while girls were given *manos*. After the ceremony, children were only carried straddling the hip. Tiesler Blos (1998) noted that a precolonial ceramic whistle depicted a child with boards on its head being carried by its mother in this position.

Tzotzil (Vogt 1993) and Tzeltal (Stross 1998) infants underwent a similar process. After being born they were cleaned, and beeswax was placed on the newborns' heads to help prevent injury from evil spirits (Stross 1998).

Gender-related tools would later be placed in the infants' hands to begin fixing the gender roles and the embodiment of social identity.

Cranial modification occurred at a time when children were vulnerable to harm because their souls were not fully anchored in their bodies and before they were fully formed social beings. Only after head binding could gender roles and destiny be fixed in precolonial Yukatek speakers. Parallels between precolonial practices and modern Yukatek and Tzotzil Maya practices strongly suggest that prior to contact with Europeans head binding was a normal part of socialization for the Maya that was important for constructing individuals who fully embodied what it was to be Maya. These ethnographic and ethnohistorical records thus suggest that we might expect every person who was a full member of Maya society to have undergone such treatment. The archaeological record, though, does not bear this out. The largest surveys of cranial modification in the Maya area have been completed by Tiesler Blos (1998, 1999), who reported that 88.65% of the crania she examined in a survey of 1,515 skulls from 94 sites in Mexico, Guatemala, and Honduras were modified (Tiesler Blos 1999). Since she also considered dental modification and did not report what percentage of the 1,515 was observable for cranial modification, it is unclear what number of skulls this actually reflected. In a more detailed report, however, 94.29% of a sample of 403 observable skulls from Mexico had some form of intentional or unintentional cultural modification (Tiesler Blos 1998). This is the highest percentage I have found in any survey or site report.

Other regional or individual site reports cited lower numbers. For example, Julie Saul and Frank Saul (1997) reported a total of 46 observable skulls from all periods from the site of Cuello in northern Belize. Of these, 20 were unmodified (43.5%), 9 were unintentionally modified (19.6%), and 17 were intentionally modified (37%). If the latter two categories are combined (as Tiesler Blos 1998 did), then the maximum percentage of modified skulls is still only 56.5%. Comparing percentages on a site-to-site basis is beyond the scope of this chapter and would likely not be constructive, given differential preservation between sites and the subjective nature of classifying skulls as modified or unmodified. The point is that meaningful percentages of unmodified skulls are turning up in the archaeological record in the Maya area. Regional studies (Tiesler Blos 1998) and local studies (Geller 2004) have failed to identify any unifying characteristics among the burials for individuals with unmodified skulls (that is, evidence of prestige). Thus the question emerges: why are there large numbers of unmodified skulls showing up in the archaeological record if cranial shaping is a basic part of growing up Maya?

Maya Cranial Modification and Soul Loss

I suggest that the way to reconcile the ethnographic and archaeological records can be found by looking at the state of children when the modification occurs. Specifically, infants are at risk of losing animating essences through their heads and are vulnerable to evil winds and thus need to be protected. The Mesoamerican notion of soul is quite different from the colloquial meaning of the word in the United States. Outlining several characteristics of the Mesoamerican soul makes this point (see Duncan 2005; Furst 1995; Geller 2004; and López Austin 1988 for reviews).

First, the soul in Mesoamerica was more of a category than a coherent unified entity. Mexica (Furst 1995) and Maya (Duncan 2005) cultures had multiple animating essences. Second, these animating essences were fully embodied, which is to say that they were found in various loci throughout the body (see Furst 1995; López Austin 1988). The important locale for this chapter is the head. Alfredo López Austin (1988) and Jill Furst (1995) note that *tonalli*, one animating essence among the Mexica, is found in the head (as discussed below). According to Stephen Houston and David Stuart (1998), for the Maya the *b'a(h)* glyph can mean "self" or "head." June Nash (1985) states that, for Tzotzil speakers, the soul leaves the body through the tongue after death.

The third characteristic of the Mesoamerican soul was that some animating essences could be diminished or possibly lost. Infants were particularly at risk for this because their souls have yet to be firmly fixed in their bodies, which rituals like the *héetz-méek'* ceremony are designed in part to do. Examining the Mexica concept of *tonalli* clarifies this point. *Tonalli* is associated with heat and destiny and resides in the head. To fortify newborns' *tonalli*, they were placed near a burning torch shortly after birth. Lighting something from that torch would have harmed the child by taking some of that essence. Similarly, the Mexica took pains to avoid stepping over children's heads and thereby injuring the *tonalli*, which would have hindered growth. Children who were sick were often described as having a loss of *tonalli* (López Austin 1988). In an effort to reduce the damage, the Mexica would let the children's hair grow long. The Mexica also interpreted a depression in the fontanelle, which can arise due to dehydration, as a result of *tonalli* loss. This was reportedly fixed by pushing up on the roof of the mouth.

It is unlikely that the Maya had any one single concept that was identical to *tonalli*. As mentioned above, however, it is clear that the Maya did have concepts of animating essences that were located in the head. Houston et al. (2006: 72) note that in the past severed heads of enemies "contained the . . . essences of the *way*, or companion spirits." Ethnographic accounts from highland Guatemala report that cutting a boy's hair prior to his first birthday would put him at risk of becoming mute or having diminished knowledge and reason (Hinojosa 1999). The Maya clearly respected the potential for soul loss, and newborns and pregnant women were most vulnerable. Modern Tzeltal speakers guard against soul loss by keeping newborns swaddled for two weeks after birth while the soul is in a tenuous place in the body (Cosminsky 2001). Also, immigrants from Guatemala and Honduras in the United States still report pushing up on the roof of the mouth of sick children in a fashion similar to that described by López Austin (1988) for the Mexica (FitzSimmons et al. 1998). Additionally, newborns are susceptible to injury from evil winds. Hence all of the doors must be closed during the birth process, and cracks in the walls must be filled with rags to prevent evil winds from entering and harming the child (Jordan 1993; Sargent and Bascope 1996). In Yucatan pregnant women and their children are protected from witches by hanging certain plants over doorways (Kunow 1996; see below).

How does this potential for soul loss and influence by evil winds relate to cranial modification? A comparison on human bodies and architecture makes the connection clear. In the Mesoamerican worldview a variety of components of the landscape were considered to be animated, not just humans, animals, and plants. Caves, mountains, houses, some ceramics, and other objects had animating essences inside them (Mock 1998). Metaphoric polysemy between humans and Maya houses suggests that covering the head of a newborn was tantamount to placing a roof on a building and that both were viewed as protecting against soul loss (Duncan and Hofling 2004). Parts of houses clearly have polysemic relations with human anatomy. Table 8.1 shows several polysemic relationships between the human body and Maya houses. The top of the roof in particular was similar to a human head. Indeed in the Codex Mendoza an Aztec lord's house is shown wearing a headdress (figure 8.1). Additionally, doorways are portrayed as mouths of houses (Plank 2004). Capstones of corbeled vaults and lintels

Figure 8.1. *Tecpan-calli* glyph ("lord place house" in Nahuatl). The portion demarcated by the bracket above the building is the headdress (figure modified from Evans 2005: figure 2.1).

Table 8.1. Associations between the Human Body and Houses in Itzaj Maya

House	Body
naj	*bäk'el*
house	(human) body
ujo'ol naj/upol naj	*jo'ol/pol*
(peak of) roof/(peak of) roof	head/head
chumuk upol naj	*tz'u' pol/chumuk pol*
center of the roof peak	crown of head/center of head
ich naj	*ich*
inside of house	face, eye
ujol naj/uchi' naj	*chi'*
doorway of the house	mouth
utaan naj	*taan*
front of house	front of body
tutzeel naj	*tzeel*
at front and back sides of house	side
näk' naj/n'ak' xa'an	*näk'*
lower part of inside wall/	belly
inside slope of roof	

Source: Modified from Duncan and Hofling (2004).

covering doorways often conflate imagery and words associated with heads and roofs (Plank 2004).

Houses also needed to be cleaned, ensouled, and protected, however, and were thus regarded as similar to newborn babies (Stross 1998; Vogt 1993). In his analysis of the fire ritual, David Stuart (1998) has noted that Maya temples were dedicated in part by having fire enter into them. This is similar to the fire drilling by which the gods animated Nahuatl babies (Furst 1995: 65) by twirling "an upright wooden piece . . . on a flat base" to create friction and initially ignite. The heat and breath used to ignite the friction helped ensoul the baby.

In an ethnographic example Brian Stross (1998) notes that, among Tzeltal speakers, upon building a house but prior to building the roof, the building would be cleaned and a chicken would be killed and placed under the roof's center beam, effectively animating the building. Colonial sources also note that houses had to be blessed prior to occupation (Cogolludo 1688 [1957]). In the twentieth century, building animation was accomplished in Yucatan in what was called the New House Ceremony (Thompson 1930). Shamans placed gourds on the corner posts of the house and offered food on a center altar. Similar ceremonies were done in Belize in the twentieth century (Thompson 1930) and among the Tzotzil Maya (Vogt 1993). The Tzotzil ceremony is important for our purposes, because of its name. The ceremony is called *hol chuk*, which literally means "binding the head of the roof" after the center or ridge pole was placed on top of the roof (Laughlin 1975). This ceremony consisted of hanging four chickens from the ridge pole of the house. The chickens were then sacrificed and cooked. The broth was offered along with liquor to each of the four corner poles of the house as well as to the center pole. A later, second rite was conducted to complete the animation of the house; but for three days the house needed to be ritually purified and taken care of, because it was weak, like a newborn.

The connection between houses and architecture is well established in the Maya area. Less explored is a point alluded to above: doorways and roofs were natural loci through which evil spirits and souls could pass. Michael David Carrasco and Kerry Hull (2002) note that capstones at various sites in the Maya area incorporate statements with the word *k'al*. This verb can mean to close or bind and is often associated with images of protection after ensoulment. More specifically, they point out that examples of certain deities, most commonly God K (K'awiil) or the Maize God, are seen moving through these capstones/portals (Carrasco and Hull 2002). God K is commonly associated with *way* transformations (Alexander 2005), which (as noted above) is a kind of companion essence found in the head. The iconography on capstones seems to be associated with the rebirth of the Maize God through a crack in the roof or through the capstone, but it mimics larger creation stories (Alexander 2005). A number of examples of associations with God K and heads (in either iconography or text) are found on Maya ceramics (Alexander 2005). God K is frequently associated with scepters, torches, and smoke. Helen Alexander (2005: 5–6) notes that "the god emerging from [the mouth of an old god] . . . does have a God K like smoking torch in his forehead." She also points out that on a number of vessels God K is portrayed as a head on the upper lip or snout of the Witz or Kawak monster. Finally, on some ceramic vases God K is portrayed as a part of the headdress of what are known as Holmul Dancers, which some researchers interpret as impersonators of the Maize God (Alexander 2005). Thus God K is associated with transformations as well as roofs, capstones, and heads. The point, for the current discussion, is that both roofs and heads are portals for the passage of animating essences. New buildings, like newborns, are in need of protection from soul loss from within or evil winds from without.

Conclusion

To return to the questions posed at the beginning of this chapter, how might we reconcile the suggestion that cranial modification was thought to be a normal part of growing up in Maya society with the evidence that a meaningful percentage of skulls found in the archaeological record show no sign of modification? I believe the answer is that binding the skull to protect against soul loss and evil winds and modifying the shape of the skull were often part of the same process, but they were not identical phenomena. Cranial modification was clearly used to create meaningful social markers, such as to index nobility or to cause some individuals to look like gods. It could be used to mark Maya from non-Maya individuals. It also may have been used, as in the northern Belize example, as a way to mark individuals on a predisposed social trajectory within Maya society, such as an occupational path. All of these suggestions have data to support them and provide insight into why cranial modification was performed among the Maya in some contexts. Early documentation of the practices may well have conflated the two processes and thereby led to overestimation of the number of modified crania that we should be finding archaeologically. Binding the head was first and foremost a way to protect against soul loss and evil winds, like placing a roof on a new building, in order to put the newborn on a path to embody what it was to be Maya. This protection was possible without modifying the shape of the skull or using the skull to mark social status and would likely have been important for all members of Maya society. Thus there is no reason to think that absence of cranial modification constitutes evidence of absence of embodiment among the Maya.

Acknowledgments

I would like to thank the editors for inviting my participation in this volume and Dr. C. Andrew Hofling for his support and assistance on this project. The anonymous reviewers for this and related publications provided valuable feedback, which improved the chapter. All remaining errors are entirely my own.

References Cited

Alexander, Helen. 2005. "God K on Maya Ceramic Vessels." Published by the Foundation Research Department of Foundation for the Advancement of Mesoamerican Studies, Inc. Accessed at: http://www.famsi.org/research/alexander/index.html.

Allison, Marvin, Enrique Gerszten, Juan Munizaga, Calogero Santoro, and Guillermo Focacci. 1981. "La práctica de la deformación craneana entre los pueblos andinos precolombinos." *Chungará* 7: 238–260.

Antón, Susan C., Carolyn R. Jaslow, and Sharon M. Swartz. 1992. "Sutural Complexity in Artificially Deformed Human (*Homo sapiens*) Crania." *Journal of Morphology* 214: 321–332.

Bennett, Kenneth A. 1965. "The Etiology and Genetics of Wormian Bones." *American Journal of Physical Anthropology* 23: 255–260.

Björk, Arne, and Lise Björk. 1964. "Artificial Deformation and Cranio-Facial Asymmetry in Ancient Peruvians." *Journal of Dental Research* 43: 353–362.

Blackwood, Beatrice, and P. M. Danby. 1955. "A Study of Artificial Cranial Deformation in New Britain." *Journal of the Royal Anthropological Institute of Great Britain and Ireland* 85: 173–191.

Bonavides Mateos, Enrique. 1992. "Rito de pasaje entre los mayas antiguos." *Estudios de Cultural Maya* 19: 397–425.

Buikstra, Jane. 1997. "Studying Maya Bioarchaeology." In *Bones of the Maya: Studies of Ancient Skeletons*, edited by S. L. Whittington and D. M. Reed, 221–228. Washington, D.C.: Smithsonian Institution Press.

Carrasco, Michael David, and Kerry Hull. 2002. "The Cosmogonic Symbolism of the Corbeled Vault in Maya Architecture." *Mexicon* 24: 26–32.

Cheverud, James M., Luci Ann P. Kohn, Lyle W. Konigsberg, and Steven R. Leigh. 1992. "Effects of Fronto-Occipital Artificial Cranial Vault Modification on the Cranial Base and Face." *American Journal of Physical Anthropology* 88: 323–345.

Cheverud, James M., and James E. Midkiff. 1992. "Effects of Fronto-Occipital Cranial Reshaping on Mandibular Form." *American Journal of Physical Anthropology* 87: 167–171.

Cogolludo, Diego López. 1688 [1957]. *Historía de Yucatán*. Colección de Grandes Crónicas Mexicanas 3. Mexico City: Editorial Academia Literaria.

Comas, Juan. 1960. "Datos para la historia de la deformación craneal en México." *Historía Mexicana* 9: 509–520.

———. 1966. *Manuel de antropología física*. Instituto de Investigaciones Históricas, Serie Antropológica 10. Mexico City: Universidad Nacional Autónoma de México.

———. 1969. "Algunos cráneos de la región maya." *Anales de Antropología* 6: 233–248.

Comas, Juan, and Paulette Marquer. 1969. *Cráneos deformados de la Isla de Sacrificios, Veracruz, México*. Instituto de Investigaciones Históricas, Cuadernos: Serie Antropológica 23. Mexico City: Universidad Nacional Autónoma de México.

Cosminsky, Sheila. 2001. "Maya Midwives of Southern Mexico and Guatemala." In *Mesoamerican Healers*, edited by B. R. Huber and A. R. Sandstrom, 179–210. Austin: University of Texas Press.

Dávalos Hurtado, Eusebio. 1945. "Tlatelolco a través de los tiempos: La deformación craneana entre los tlatelolca." *Memorias de la Academia Mexicana de la Historía Correspondiente de la Real de Madrid* 4: 111–130.

———. 1946. "Las deformaciones craneanas." In *México prehispánico: Culturas, deidades, monumentos*, edited by J. A. Vivó, 831–840. Mexico City: Editorial E. Hurtado.

———. 1965. *Temas de antropología física*. Mexico City: Instituto Nacional de Antropología e Historia, Secretaría de Educación Pública.

Dean, Valerie L. 1995. "Sinus and Meningeal Vessel Pattern Changes Induced by Artificial Cranial Deformation." *International Journal of Osteoarchaeology* 5: 1–14.

Demarest, Arthur A., Prudence M. Rice, and Don S. Rice, editors. 2004. *The Terminal Classic in the Maya Lowlands: Collapse, Transition, and Transformation*. Boulder: University Press of Colorado.

Dembo, Adolfo, and José Imbelloni. 1938. *Deformaciones intencionales del cuerpo humano de carácter étnico*. Buenos Aires: J. Anesi.

Dingwall, Eric. 1931. *Artificial Cranial Deformation: A Contribution to the Study of Ethnic Mutilations*. London: J. Bale Sons and Danielsson.

Duncan, William N. 2005. "Ritual Violence among the Postclassic Maya: A Bioarchaeological Analysis." Doctoral dissertation, Southern Illinois University, Carbondale, Department of Anthropology.

Duncan, William N., and C. Andrew Hofling. 2004. "Polysemic Association between Cranial Modification and Architectural Construction." Presented in the symposium "Tensions, Theory, and Directions in Bioarchaeology" at the 103rd Annual Meeting of the American Anthropological Association, Atlanta, Ga.

Evans, Susan Toby. 2005. "The Aztec Palace under Spanish Rule: Disk Motifs in the Maya de Mexico de 1550 (Uppsala Map or Mapa de Santa Cruz)." In *The Postclassic to Spanish-Era Transition in Mesoamerica*, edited by S. Kepecs and R. T. Alexander, 13–35. Albuquerque: University of New Mexico Press.

Falkenburger, Frédéric. 1938. "Recherches anthropologiques sur la déformation artificielle du crâne." *Journal de la Société des Américanistes* 30: 1–69.

Feindel, William. 1988. "Cranial Clues to the Mysterious Decline of the Maya Civilization: The Hippocampal Hypothesis." *América Indígena* 48: 215–219.

FitzSimmons, Ellen, Jack H. Prost, and Sharon Peniston. 1998. "Infant Head Molding: A Cultural Practice." *Archives of Family Medicine* 7: 88–90.

Flower, William H. 1882. "On a Collection of Monumental Heads and Artificially Deformed Crania from the Island of Mallicollo, in the New Hebrides." *Journal of the Anthropological Institute of Great Britain and Ireland* 11: 75–81.

Frieß, Martin, and Michel Baylac. 2003. "Exploring Artificial Cranial Deformation Using Elliptic Fourier Analysis of Procrustes Aligned Outlines." *American Journal of Physical Anthropology* 122: 11–22.

Furst, Jill Leslie McKeever. 1995. *The Natural History of the Soul in Ancient Mexico*. New Haven, Conn.: Yale University Press.

Geller, Pamela. 2004. "Transforming Bodies, Transforming Identities: A Consideration of Pre-Columbian Maya Corporeal Beliefs and Practices." Doctoral dissertation, University of Pennsylvania, Department of Anthropology.

Gerszten, Peter C. 1993. "An Investigation into the Practice of Cranial Deformation among the Pre-Columbian Peoples of Northern Chile." *International Journal of Osteoarchaeology* 3: 87–89.

Gerszten, Peter C., and Enrique Gerstzen. 1995. "Intentional Cranial Deformation: A Disappearing Form of Self-Mutilation." *Neurosurgery* 37: 374–382.

Goodrich, James T., and Matteo Tutino. 2001. "An Annotated History of Craniofacial Surgery and Intentional Cranial Deformation." *Neurosurgery Clinics of North America* 12: 45–68.

Gottlieb, Karen. 1978. "Artificial Cranial Deformation and the Increased Complexity of the Lambdoid Suture." *American Journal of Physical Anthropology* 48: 213–314.

Hinojosa, Servando. 1999. "Spiritual Embodiment in a Highland Maya Community." Doctoral dissertation, Tulane University, Department of Anthropology.

Hippocrates. 400 BC [1952]. "On Airs, Waters, and Places." In *Hippocratic Writings*, translated by F. Adams, 9–18. *Encyclopaedia Britannica*, vol. 10. Chicago: Encyclopaedia Britannica.

Houston, Stephen, and David Stuart. 1998. "The Ancient Maya Self: Personhood and Portraiture in the Classic Period." *Res: Anthropology and Aesthetics* 33: 73–101.

Houston, Stephen, David Stuart, and Karl Taube. 2006. *The Memory of Bones: Body, Being, and Experience among the Classic Maya*. Austin: University of Texas Press.

Jordan, Brigitte. 1993. *Birth in Four Cultures*. 4th ed. Revised and expanded by Robbie Davis-Floyd. Prospect Heights, Ill.: Waveland Press.

Joyce, Rosemary. 2000. "Girling the Girl and Boying the Boy: The Production of Adulthood in Ancient Mesoamerica." *World Archaeology* 31: 473–483.

———. 2001. "Negotiating Sex and Gender in Classic Maya Society." In *Gender in Pre-

Hispanic America, edited by C. F. Klein, 109–141. Washington, D.C.: Dumbarton Oaks Research Library and Collection.

Kohn, Luci Ann P., Steven R. Leigh, Susan C. Jacobs, and James M. Cheverud. 1993. "Effects of Annular Cranial Vault Modification on the Cranial Base and Face." *American Journal of Physical Anthropology* 90: 147–168.

Konigsberg, Lyle W., Luci Ann P. Kohn, and James M. Cheverud. 1993. "Cranial Deformation and Nonmetric Trait Variation." *American Journal of Physical Anthropology* 90: 35–48.

Kunow, Marianna A. 1996. "Curing and Curers in Pisté, Yucatan, Mexico." Doctoral dissertation, Tulane University, Department of Anthropology.

Laughlin, Robert M. 1975. *The Great Tzotzil Dictionary of San Lorenzo Zinacantán*. Smithsonian Contributions to Anthropology No. 19. Washington, D.C.: Smithsonian Institution Press.

López Austin, Alfredo. 1988. *The Human Body and Ideology: Concepts of the Ancient Nahuas*. Salt Lake City: University of Utah Press.

Magitot, M. 1885. "Essai sur les mutilations ethniques." *Bulletins de la Société d'Anthropologie de Paris* 8: 21–25.

McGibbon, Walter. 1912. "Artificially Deformed Skulls with Special Reference to the Temporal Bone and Its Tympanic Portion." *Laryngoscope* 22: 1165–1184.

McNeill, R. William, and George N. Newton. 1965. "Cranial Base Morphology in Association with Intentional Cranial Vault Deformation." *American Journal of Physical Anthropology* 23: 241–253.

Mock, Shirley. 1998. *The Sowing and the Dawning: Termination, Dedication and Transformation in the Archaeological and Ethnographic Record of Mesoamerica*. Albuquerque: University of New Mexico Press.

Moss, Melvin L. 1958. "The Pathogenesis of Artificial Cranial Deformation." *American Journal of Physical Anthropology* 16: 269–286.

Nash, June. 1985. *In the Eyes of the Ancestors: Belief and Behavior in a Maya Community*. Prospect Heights, Ill.: Waveland Press.

Oetteking, Bruno. 1924. "Declination of the Pars Basilaris in Normal and in Artificially Deformed Skulls." *Indian Notes and Monographs* 27: 1–25.

O'Loughlin, Valerie D. 1996. "Comparative Endocranial Vascular Changes Due to Craniosynostosis and Artificial Cranial Deformation." *American Journal of Physical Anthropology* 101: 369–385.

———. 2004. "Effects of Different Kinds of Cranial Deformation on the Incidence of Wormian Bones." *American Journal of Physical Anthropology* 123: 146–155.

Ossenberg, Nancy S. 1970. "The Influence of Artificial Cranial Deformation on Discontinuous Morphological Traits." *American Journal of Physical Anthropology* 33: 357–371.

Plank, Shannon. 2004. *Maya Dwellings in Hieroglyphs and Archaeology: An Integrative Approach to Ancient Architecture and Spatial Cognition*. BAR International Series 1324. Oxford: Archaeopress.

Redfield, Robert, and Alfonso Villa Rojas. 1934. *Chan Kom: A Maya Village*. Publication 448. Washington, D.C.: Carnegie Institution of Washington.

Rogers, Spencer L. 1975. *Artificial Deformation of the Head: New World Examples of Ethnic Mutilation and Notes on Its Consequences*. San Diego Museum Papers No. 8. San Diego: San Diego Museum of Man.

Romano Pacheco, Arturo. 1974. "Deformación cefálica intencional." In *Antropología física: Época prehispánica 3*, edited by J. Romero, 195–227. Mexico City: Colección México, Panorama Histórico y Cultural, Instituto Nacional de Antropología e Historia.

Romero Molina, Javier. 1970. "Dental Mutilation, Trephination, and Cranial Deformation." In *Handbook of Middle American Indians*, vol. 9, edited by T. D. Stewart, 50–67. Austin: University of Texas Press.

Sargent, Carolyn, and Grace Bascope. 1996. "Ways of Knowing about Birth in Three Cultures." *Medical Anthropology Quarterly*, new series 10: 213–236.

Saul, Julie M., and Frank P. Saul. 1997. "The Preclassic Skeletons from Cuello." In *Bones of the Maya: Studies of Ancient Skeletons*, edited by S. L. Whittington and D. M. Reed, 28–50. Washington, D.C.: Smithsonian Institution Press.

Stross, Brian. 1998. "Seven Ingredients in Mesoamerican Ensoulment, Dedication and Termination in Tenejapa." In *The Sowing and the Dawning: Termination, Dedication, and Transformation in the Archaeological and Ethnographic Record of Mesoamerica*, edited by S. Boteler Mock, 31–46. Albuquerque: University of New Mexico Press.

Stuart, David. 1998. "'The Fire Enters His House': Architecture and Ritual in Classic Maya Texts." In *Function and Meaning in Classic Maya Architecture*, edited by S. D. Houston, 479–517. Washington, D.C.: Dumbarton Oaks Research Library and Collection.

Thompson, J. Eric S. 1930. *Ethnology of the Mayas of Southern and Central British Honduras*. Publication 274, Anthropology Series, vol. 17, no. 2. Chicago: Field Museum of Natural History.

Tiesler Blos, Vera. 1995. "La deformación cefálica entre los mayas: Aspectos neurofisiológicos." *Memorias del Segundo Congreso Internacional de Mayistas* 2: 662–679.

———. 1998. *La costumbre de la deformación cefálica entre los antiguos mayas: Aspectos morfológicos y culturales*. Mexico City: INAH.

———. 1999. "Head Shaping and Dental Decoration among the Ancient Maya: Archaeological and Cultural Aspects." Paper presented at the 64th Meeting of the Society of American Archaeology, Chicago.

Tommaseo, Mila, and Andrea Drusini. 1984. "Physical Anthropology of Two Tribal Groups in Amazonia Peru (with Reference to Artificial Cranial Deformation)." *Zeitschrift für Morphologie und Anthropologie* 74: 315–333.

Topinard, Paul. 1879. "Moulages d'un crâne macrocéphale et d'un crâne de l'époque de la Pierre Polie." *Société d'Anthropologie de Paris Bulletin* 3: 116–121.

Torquemada, Juan de. 1613 [1943]. *Monarquía indiana*. 3rd ed. 3 vols. Mexico City: Editorial Salvador Chavez Hayhoe.

Torres-Rouff, Christina. 2002. "Cranial Vault Modification and Ethnicity in Middle Horizon San Pedro de Atacama, Chile." *Current Anthropology* 43: 163–171.

———. 2003. "Shaping Identity: Cranial Vault Modification in the Pre-Columbian Andes." Doctoral dissertation, University of California, Santa Barbara, Department of Anthropology.

Tozzer, Alfred M. 1941. *Landa's Relación de las Cosas de Yucatan: A Translation Edited with Notes.* Cambridge, Mass.: Peabody Ethnology, Harvard University.

Tubbs, R., E. George Salter, and W. Jerry Oakes. 2006. "Artificial Deformation of the Human Skull: A Review." *Clinical Anatomy* 19: 372–377.

Villa Rojas, Alfonso. 1945. *The Maya of East Central Quintana Roo.* Publication 559. Washington, D.C.: Carnegie Institution of Washington.

Vogt, Evon Z. 1993. *Tortillas for the Gods: A Symbolic Analysis of Zinacanteco Rituals.* Norman: University of Oklahoma Press.

White, Christine D. 1996. "Sutural Effects of Fronto-Occipital Cranial Modification." *American Journal of Physical Anthropology* 100: 397–410.

9

The Complex Relationship between Tiwanaku Mortuary Identity and Geographic Origin in the South Central Andes

KELLY J. KNUDSON AND DEBORAH E. BLOM

When faced with the political integration and expansion of states and empires, incorporated peoples often create and manipulate political, social, and religious identities in their interactions with larger and more powerful polities. In the Andes, as in other regions, states and empires like the Tiwanaku (ca. AD 500–1100) and the Inka (AD 1400–1532) integrated large geographical areas through political, social, and economic strategies, such as colonization, forced migration and resettlement, ideological manipulation, trade, and taxation (e.g., Bauer 1996, 2004; Bauer and Stanish 2001; D'Altroy 1992, 2002; D'Altroy and Schreiber 2004; Kolata 1993, 1996, 2003; Stanish 2003). In this chapter we present a bioarchaeological approach to the study of identity creation and manipulation to elucidate the complex roles of both heartland and hinterland populations in the process of Tiwanaku political integration and expansion.

More specifically, we utilize multiple lines of bioarchaeological and archaeological evidence to examine the nature of Tiwanaku influence and changes in Tiwanaku mortuary identities between AD 500 and 1100. During this period, called the Middle Horizon, the Tiwanaku polity exerted influence far beyond its heartland in the Lake Titicaca Basin to affect Tiwanaku-affiliated sites in the coastal Moquegua Valley of southern Peru (like Chen Chen) and in northern Chile at the San Pedro de Atacama oasis, among others. Using multiple lines of evidence, including isotopic signatures, biodistance analysis, cranial modification styles, mortuary artifacts, and burial treatments, we argue that at least some individuals buried at Chen Chen were in fact immigrants from the Tiwanaku heartland and that the Moquegua settlements articulated with the Tiwanaku polity as a colony or diaspora. Hence there is a clear correlation between Tiwanaku mortuary traditions and biological affiliation with the Tiwanaku heartland for the individuals buried in the cemetery of Chen Chen. In contrast, in the northern

Chilean oasis of San Pedro de Atacama, the Tiwanaku mortuary artifacts were buried with a local population, not immigrants from the Lake Titicaca Basin. In San Pedro de Atacama, local inhabitants manipulated their ethnic, social, political, and/or religious identities as they articulated with the distant Tiwanaku polity. Through this example of identity formation and manipulation in the archaeological record, we demonstrate the potential of multiple lines of bioarchaeological evidence to elucidate the complex relationships of material culture, geographic origin, and identity.

Identity Manipulation and Bioarchaeology

For over 50 years scholars have conceived of identity, including ethnic identity, as situational, flexible, and dynamic (e.g., Barth 1969; Díaz-Andreu et al. 2005; Hutchinson and Smith 1996; Jones 1997; Romanucci-Ross and De Vos 1995). Given the time-depth of the archaeological record, however, much archaeological research has focused on investigating long-standing identities through consumption patterns and material culture (e.g., Aldenderfer 1993; McGuire 1982; Pyszczyk 1989; Rattray 1990; Reycraft 2005; Smith 2003). When ethnographic and historical evidence is taken into account, a more dynamic picture, with more evidence of the manipulation and mediation of different identities, emerges (e.g., Bentley 1987; Bowser 2000; Epperson 1999; Gosden 2004; Larick 1991; Lightfoot 1995; Rothschild 2006; Stahl 1991).

Likewise, bioarchaeology can contribute much to our understanding of identity manipulation in the archaeological record by combining different lines of evidence on the relationships among mortuary patterns, genetic affiliation, ethnicity, gender, and other social and religious factors and identities. For example, the genetic affiliation of an individual is static and cannot be manipulated. Material culture used by an individual during her or his life can be changed regularly, however, and can passively or actively communicate information through style and other attributes (e.g., Bourdieu 1990; Conkey and Hastorf 1990; Dietler and Herbich 1998; Hegmon 1998; Sackett 1977; Wells 1998; Wiessner 1983). In contrast, cranial modification styles reflect family or community decisions, and likely group identity, since an individual's cranium must be modified over the first few years of life and cannot be changed or manipulated later in life (e.g., Blom 2005b; Torres-Rouff 2002, this volume). Similarly, while mortuary practices may reflect the individual as she or he was in life, the mortuary activities themselves reflect the actions of the community and/or family who buried that individual (e.g., Arnold and Wicker 2001; Brown 1971; Carr 1995; Chesson 2001; Dillehay 1995; Gillespie 2001; Rakita et al. 2005; Saxe 1970; Silverman and Small 2002).

By using various lines of bioarchaeological evidence to elucidate different identities, we can address key questions in the South Central Andes. For example, how did individuals living in dispersed settlements throughout the Tiwanaku realm differentially manipulate their identities when articulating with a more powerful foreign polity? How did individuals passively or actively use material culture to define themselves in relation to other individuals and to define themselves in relation to other social groups? How do individuals use material culture to manipulate and communicate meanings and information and advertise their identities?

Tiwanaku Mortuary Identity and Geographic Origin at Chen Chen, Peru

Tiwanaku Presence in the Moquegua Valley of Southern Peru

The heartland of the Tiwanaku polity, including the capital of Tiwanaku, is located in the southeastern Lake Titicaca Basin of Bolivia. From the large urban center of Tiwanaku, the Tiwanaku polity likely exerted political, economic, and religious control over much of the Lake Titicaca Basin during the Middle Horizon period (ca. AD 500–1100). However, Tiwanaku-style material culture, and presumably Tiwanaku influence, is spread throughout the South Central Andes. In southern Peru, the Moquegua Valley is home to the three large Tiwanaku-affiliated site complexes of Chen Chen, Omo, and Río Muerto and at least 30 smaller Tiwanaku-affiliated sites (figures 9.1, 9.2) (Goldstein 2005). Based on residential and ritual architecture and artifact assemblages, with ceramics indistinguishable from *altiplano* counterparts, numerous scholars have argued that the Moquegua Valley was home to Tiwanaku colonies or diaspora communities that provided the Tiwanaku heartland with ritually important low-altitude crops such as maize (Goldstein 1993, 2005; Kolata 1993; Mujica et al. 1983; Owen 2005).

Multiple Lines of Evidence for Identity at Chen Chen

The cemetery at Chen Chen, which is the largest Tiwanaku-affiliated mortuary complex in the South Central Andes, provides a unique opportunity to investigate Tiwanaku identities and the relationship between the Tiwanaku heartland and the Moquegua Valley. Here we investigate biological affilia-

Above: Figure 9.1. The Tiwanaku-affiliated site complex of Chen Chen (photo by Deborah E. Blom).

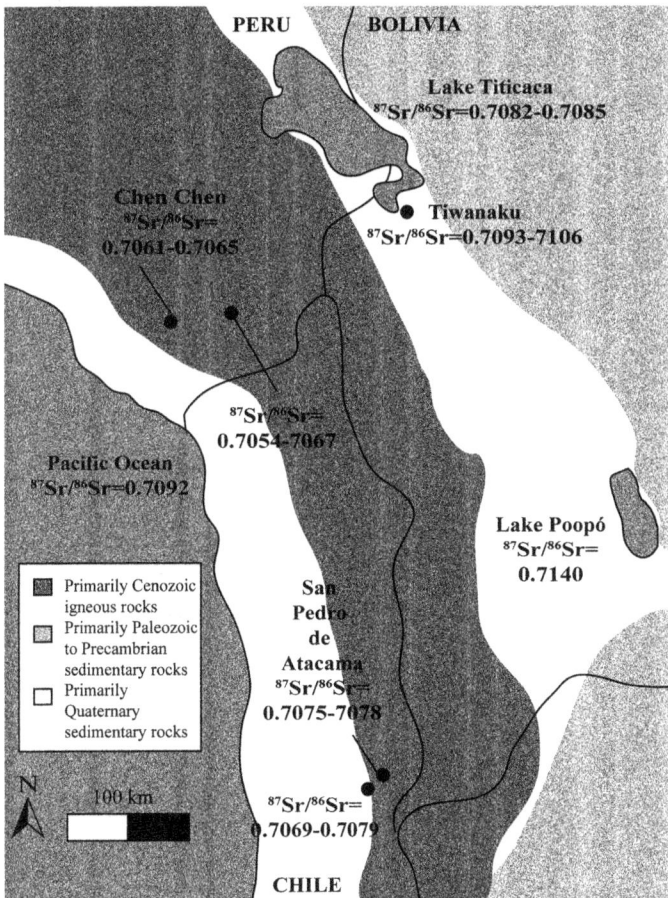

Left: Figure 9.2. Geologic sketch map of the South Central Andes, showing the sites discussed in the text. Strontium isotope ratios measured in exposed bedrock, groundwater, and modern and archaeological small mammals in the region are also displayed (Grove et al. 2003; Hawkesworth et al. 1982; James 1982; Knudson and Price 2007; Knudson et al. 2004; Knudson et al. 2005; Rogers and Hawkesworth 1989).

tion through genetic analyses and geographic origins through strontium isotope analysis. We then compare those data to cranial modification data, which have been linked to group identity in Tiwanaku society (Blom 2005a, 2005b; Blom and Janusek 2004; Hoshower et al. 1995), and mortuary patterns, such as tomb type, burial orientation, and mortuary artifacts, to examine the relationship between mortuary and other cultural traditions and biological origin.

The genetic affiliation of individuals buried at Chen Chen was first investigated through biodistance analysis of cranial nonmetric traits (Blom 2005a; Blom et al. 1998) and then by using mitochondrial DNA (Lewis et al. 2007; Lewis and Stone 2005). These data indicate that the mortuary populations at Chen Chen and Tiwanaku are genetically closely related. Since the genetic similarity may result from individuals moving from Chen Chen to Tiwanaku or vice versa, however, we then used strontium isotope analysis of archaeological human enamel and bone samples from Chen Chen to investigate geographic origins (Knudson and Price 2007; Knudson et al. 2004). Briefly, strontium isotope ratios vary according to bedrock geology and can be used to differentiate individuals who spent the years of enamel and/or bone formation in the southeastern Lake Titicaca Basin or the Moquegua Valley (figure 9.2) (Bentley 2006; Price et al. 1994). Based on enamel strontium isotope ratios, at least 2 females out of the 25 individuals sampled from Chen Chen likely lived in the southeastern Lake Titicaca Basin during the first three years of their lives yet migrated to and were buried at Chen Chen sometime during adulthood (figure 9.3) (Knudson and Price 2007; Knudson et al. 2004). The relatively small number of documented first-generation migrants from the southeastern Lake Titicaca Basin, however, may imply that the Tiwanaku-affiliated sites in the Moquegua Valley were not populated by large numbers of first-generation migrants and that the community was a long-lived and self-sufficient colony with biological ties to the Tiwanaku heartland. In conclusion, multiple lines of evidence demonstrate that the biological identity of the individuals buried at Chen Chen was in fact derived from the Tiwanaku heartland.

First-generation *altiplano* immigrants at Chen Chen, however, cannot be distinguished based on cranial modification style, mortuary artifacts, tomb architecture, or tomb placement. All burials analyzed from the cemetery of Chen Chen are associated with Chen Chen–style ceramics, which are visually indistinguishable from *altiplano* Tiwanaku ceramics (Goldstein 2005), so artifacts do not distinguish the nonlocals from the locals buried at Chen Chen. In addition, the individuals buried at Chen Chen who have

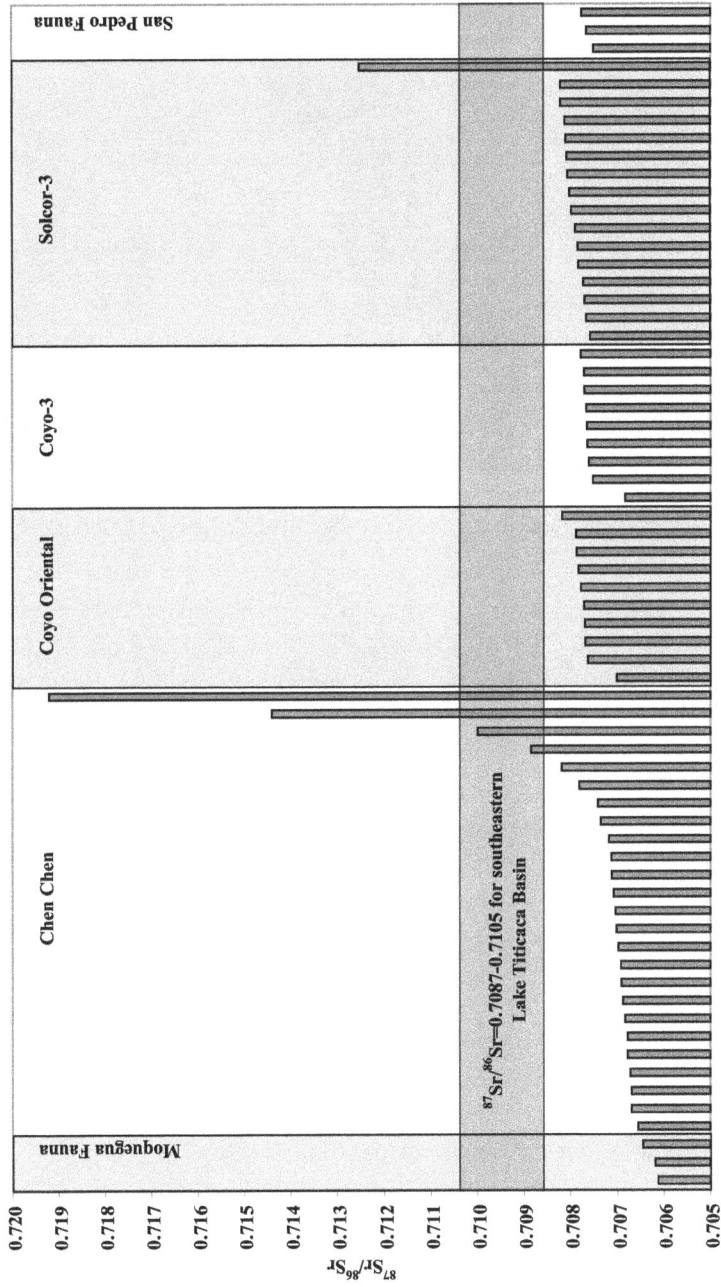

Figure 9.3. Strontium isotope ratios from archaeological human tooth enamel from individuals buried at the sites of Chen Chen, Coyo Oriental, Coyo-3, and Solcor-3. The following local ranges for each study region were determined by taking the mean strontium isotope ratios from archaeological and modern small mammal bones plus and minus two standard deviations: $^{87}Sr/^{86}Sr = 0.7087–0.7105$ for the southeastern Lake Titicaca Basin, $^{87}Sr/^{86}Sr = 0.7074–0.7079$ for the San Pedro de Atacama region, $^{87}Sr/^{86}Sr = 0.7059–0.7066$ for the Moquegua Valley, and $^{87}Sr/^{86}Sr = 0.7058–0.7082$ for the Ilo Valley.

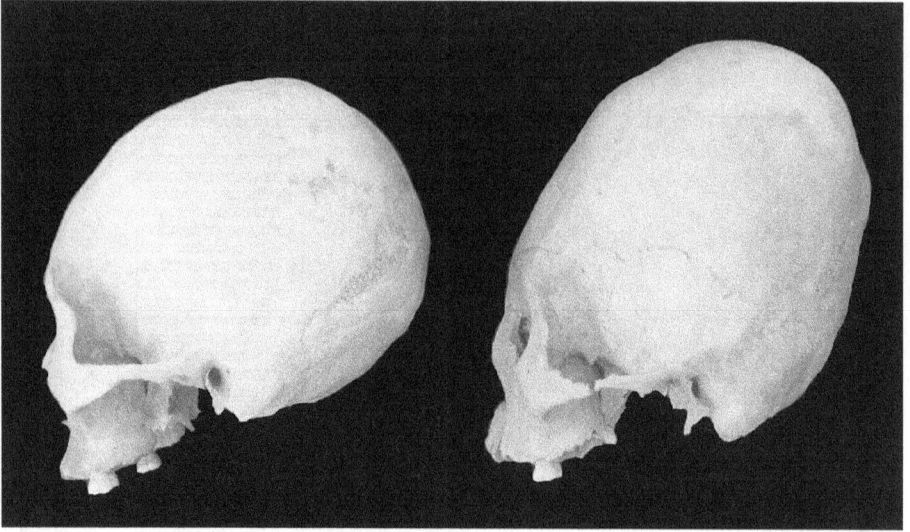

Figure 9.4. Fronto-occipital cranial shape modification in specimen number M1-0169 compared with the unmodified crania of specimen number M1-2068 (photo by Deborah E. Blom).

southern Lake Titicaca Basin signatures in their tooth enamel or bone (M1-3840, $^{87}Sr/^{86}Sr = 0.708843$; M1-S/NB092, $^{87}Sr/^{86}Sr = 0.709995$; and M1-1600, $^{87}Sr/^{86}Sr = 0.708850$) all exhibit fronto-occipital cranial modification (figure 9.4). But this type of cranial modification does not distinguish these individuals from other *local* individuals buried at Chen Chen either. Of 287 crania at Chen Chen, 200 exhibited fronto-occipital modification (Blom 1999). Subtle differences in cranial modification did not vary with tomb type or tomb location at Chen Chen (Blom 1999) and cannot be used to identify immigrants from the Lake Titicaca Basin identified through strontium isotope analysis.

Therefore, our data show that at Chen Chen identification with the Tiwanaku polity was not confined to individuals born in the Lake Titicaca Basin and did not distinguish between different individuals or different groups buried at Chen Chen. In this case, the material culture, cranial modification data, and strontium isotope data all point to a community-wide Tiwanaku social and political identity and perhaps ethnic identity.

Identity Manipulation and Tiwanaku Mortuary Artifacts in San Pedro de Atacama, Chile

Tiwanaku Influence in San Pedro de Atacama

The San Pedro de Atacama region of northern Chile is a large oasis in the Atacama Desert (figure 9.2). Because of excellent preservation and extensive cemetery excavations in the oasis, researchers have elucidated a long cultural sequence, including the presence of Tiwanaku-style artifacts during the Middle Horizon period (ca. AD 500–1100). In the San Pedro de Atacama oasis, the mortuary artifacts are characterized by the presence of Tiwanaku-style textiles and wooden snuff tablets, tubes, and spatulas used for the inhalation of hallucinogenic snuffs, with fewer quantities of Tiwanaku-style pyroengraved bone tubes and vessels, ceramics, and gold *keros*, which are vessels presumably for the consumption of maize beer (Bravo and Llagostera 1986; Costa-Junqueira and Llagostera 1994; LePaige 1961; Llagostera et al. 1988; Oakland Rodman 1992; Stovel 2001; Torres 1985, 1987; Torres et al. 1991). Based on the presence of Tiwanaku-style mortuary artifacts as well as biodistance analysis of cranial nonmetric traits and cranial modification studies, a number of scholars have argued that a population of immigrants from the Tiwanaku heartland in the Lake Titicaca Basin lived in San Pedro de Atacama (Kolata 1993; Oakland Rodman 1992; Varela and Cocilovo 2000).

Material Culture and Geographic Origins in San Pedro de Atacama

Strontium isotope analysis of 53 individuals buried with Tiwanaku-style and/or local-style mortuary artifacts in the San Pedro de Atacama cemeteries of Coyo Oriental, Coyo-3, and Solcor-3 has not identified any individuals whose enamel strontium isotope signatures are from the southeastern Lake Titicaca Basin (figure 9.3) (Knudson 2004; Knudson and Price 2007). Therefore, no immigrants from the Tiwanaku heartland have been identified in Middle Horizon San Pedro de Atacama cemeteries. Other lines of evidence, notably cranial modification studies and biodistance analysis of cranial nonmetric traits at Solcor-3, also support the hypothesis that local inhabitants of San Pedro de Atacama adopted Tiwanaku-style mortuary artifacts without biological ties to the Tiwanaku heartland (Llagostera et al. 1988; Torres-Rouff 2002). In addition, the Tiwanaku-style mortuary artifacts in San Pedro de Atacama coexist with artifacts such as ceramics, baskets, and textiles in the local styles. This is significantly different from the Moquegua settlements, where local styles are absent (Goldstein 2005).

The Role of Tiwanaku-Style Mortuary Artifacts in San Pedro de Atacama

If the San Pedro de Atacama oasis was not populated by first-generation migrants from the Tiwanaku heartland, how can we explain the presence of Tiwanaku mortuary artifacts in San Pedro de Atacama? Numerous scholars have explored the economic influence between the Tiwanaku polity and the inhabitants of San Pedro de Atacama (e.g., Kolata 1993; Torres and Conklin 1995). San Pedro de Atacama was likely an important oasis on caravan routes and may also have been a source for salt, copper, lapis lazuli, and turquoise (Stanish 2003; Torres and Conklin 1995).

The trade of Tiwanaku-style artifacts for food, water, salt, and raw materials from the oasis, however, does not adequately explain the preponderance of Tiwanaku-style ritual goods like snuffing paraphernalia in San Pedro de Atacama cemeteries (Torres and Conklin 1995). Did inhabitants of the San Pedro de Atacama oasis adopt Tiwanaku religious identity during the Middle Horizon? The presence of Tiwanaku-style snuff tablets in cemeteries where some individuals were buried with snuff tablets carved in local styles may be the result of some individuals adopting and advertising Tiwanaku belief systems. Local San Pedro de Atacama inhabitants may have gained power or prestige through their use of Tiwanaku-style snuffing paraphernalia instead of snuff tablets and tubes in local styles; for example, Agustín Llagostera (1996) argues that local elites gained power in San Pedro de Atacama through their association with nonlocal artifacts (see also Helms 1993). Individuals buried with Tiwanaku-style mortuary artifacts are in fact associated with high-status goods like metal artifacts and greater numbers of artifacts (e.g., Bravo and Llagostera 1986; Costa-Junqueira and Llagostera 1994; Llagostera et al. 1988; Oakland Rodman 1992). While status differences do exist in the San Pedro de Atacama cemeteries, the high percentages of individuals identified as "high-status" imply that the inclusion of Tiwanaku-style artifacts was not limited to a few, elite individuals.

In addition, while Tiwanaku-style iconography exists on some snuff tablets, ritual acts such as the act of snuffing itself presumably did not change depending on the style of the snuffing paraphernalia. Therefore, while Tiwanaku-style ritual objects were important mortuary artifacts in San Pedro de Atacama, it does not appear that the adoption of Tiwanaku-style snuffing paraphernalia was accompanied by a dramatic change in ritual activity in the oasis. In fact, this type of drug paraphernalia is rare in Moquegua, where abundant evidence of Tiwanaku ritual behavior is present (Goldstein 2005).

While specific types of Tiwanaku-style mortuary artifacts were incorporated into San Pedro de Atacama cemeteries, the inhabitants of San Pedro de Atacama do not appear to have adopted Tiwanaku identity as expressed in ritual or mortuary contexts. As previously discussed, Tiwanaku mortuary traditions in the Moquegua Valley site of Chen Chen are very different from those in the San Pedro de Atacama oasis. While some individuals are buried with Tiwanaku-style mortuary artifacts, burial treatment, tomb type, and burial orientation are all characteristic of local Atacameño practices and mortuary identity (e.g., Costa-Junqueira and Llagostera 1994; Llagostera et al. 1988; Torres and Conklin 1995).

Identity Manipulation in San Pedro de Atacama

Alternatively, scholars such as Amy Oakland Rodman (1992: 316) argued that "the oasis was home to a foreign *altiplano* population who maintained for centuries an ethnic identity visible in a distinct textile style." Oakland Rodman made an excellent point that clothing and textile style are more indicative of ethnic identity than other types of artifacts. Given this, it is clear that some individuals buried in the cemetery of Coyo Oriental were actively displaying Tiwanaku affiliation in their mortuary practices, and Oakland Rodman argued that these individuals derived from the *altiplano*. Knudson's research, however, indicates that the situation was more complex. Surprisingly, none of the individuals included in both Oakland Rodman's textile analysis and strontium isotope studies were migrants from the *altiplano* (figure 9.3). In light of this, a closer examination of grave inclusions reveals other discrepancies at odds with the migration hypothesis. The individuals buried in Tiwanaku-style textiles such as tunics and headdresses were not necessarily the individuals buried with other Tiwanaku-style artifacts, and some individuals buried with Tiwanaku-style snuff tablets were not buried with Tiwanaku-style textiles, like individual CO-4093 from Coyo Oriental (Oakland Rodman 1992; Torres 1987).

When compared with sites where ethnic enclaves have been identified, the archaeological evidence from San Pedro de Atacama cemeteries shows key differences. For example, distinct enclaves of individuals whose ethnic identity was different from that of their neighbors can be found at sites such as Teotihuacan in central Mexico (Price et al. 2000; Rattray 1990; White, Spence, et al. 2004; White, Storey, et al. 2004) and Hacinebi in Turkey (Stein 1999, 2002). In these ethnic enclaves, mortuary patterns are consistent with geographic origins, residential architecture, and domestic artifact styles. In the San Pedro de Atacama oasis, however, ethnic identity as recognized through textile styles is not consistent with mortuary traditions, cranial modification styles, genetic relationships, or geographic origins. Since the Tiwanaku presence in the San Pedro de Atacama oasis does not appear to be associated with coercion (Costa-Junqueira et al. 1998; Neves et al. 1999; Neves and Costa 1998) or population movement (Knudson and Price 2007), the oasis inhabitants must have actively pursued the manipulation of Tiwanaku ethnic identity.

Discussion and Conclusion

We have argued that bioarchaeology can address a number of important questions regarding identity formation and manipulation. How did individuals living in various settlements throughout the Tiwanaku realm differentially manipulate their identities when articulating with a more powerful foreign polity? How did individuals passively or actively use material culture to define themselves in relation to other individuals and to define themselves in relation to other social groups? How do individuals use material culture to manipulate and communicate meanings and information and advertise their identities?

At Chen Chen in southern Peru, homogeneity in material culture, mortuary patterns, and cranial modification styles coexists with heterogeneity in enamel strontium isotope signatures. This implies that first-generation migrants from the *altiplano* were not identified by themselves or by others in their community as distinct based on mortuary artifacts, tomb type, or cranial modification. Individuals living at Chen Chen appear to have retained or reinvented an identity with their homeland over multiple generations. In the San Pedro de Atacama oasis, however, heterogeneity in mortuary artifacts and the inclusion of Tiwanaku-style artifacts in some tombs at the cemeteries of Coyo Oriental, Coyo-3, and Solcor-3 is in contrast to the homogeneity in enamel strontium isotope ratios and geographic origins. In these cemeteries, some individuals utilized mortuary artifacts to advertise Tiwanaku affiliation. It is possible that the manipulation of identity in the oasis resulted from political, economic, and religious interactions with the Tiwanaku polity. But the presence of ritual objects and mortuary textiles in Tiwanaku styles does not point to a purely political or economic relationship between the two regions. This is an example of individuals manipulating their mortuary identity, and possibly ethnic identity, as they articulated with but were not dominated by (e.g., Rothschild 2006) the larger and more powerful Tiwanaku polity to the northeast.

In conclusion, we have used a number of different lines of bioarchaeological and archaeological evidence to investigate identities in the Middle Horizon South Central Andes. Individuals in the Moquegua Valley and in the San Pedro de Atacama oasis utilized and manipulated Tiwanaku identity in a variety of ways, depending on their economic, political, and/or religious relationship with the Tiwanaku polity. Through contextualized bioarchaeological analyses of identity at the level of the individual, we have elucidated the complexity of relationships in the South Central Andes during the expansion of the Tiwanaku polity.

Acknowledgments

Our work over the years would not have been possible without the generous financial support of the National Science Foundation (BCS-0202329 and SBR-9708001), the Wenner-Gren Foundation (grant number 5863), the Latin American Studies Department at the University of Wisconsin at Madison, and the Geological Society of America. In addition, archaeological human remains from Tiwanaku were excavated under the auspices of the Proyecto Wila Jawira, which was supported by multiple grants awarded to Alan L. Kolata (National Science Foundation BNS-8607541, BNS-8805490, DEB-9212641; National Oceanic and Atmospheric Administration GC-95-174; National Endowment for the Humanities RO-21806-88, RO-21368-86). The following individuals and institutions generously provided contextual information, laboratory and museum access, and/or logistical support: James H. Burton, Centro de Investigaciones Arqueológicas de Arequipa, José Cocilovo, María Antonietta Costa-Junqueira, Nicole Couture, Paul D. Fullagar, Agustín Llagostera, Museo Contisuyo, Amy Oakland Rodman, T. Douglas Price, Proyecto Arqueológico Pumapunku-Akapana, Paula Tomczak, Christina Torres-Rouff, Hugo Varela, and Bertha Vargas. Finally, we began to examine the issues discussed in this chapter when we were invited to participate in a symposium entitled "Tensions, Theory, and Directions in Bioarchaeology," organized by Kenneth C. Nystrom at the 103rd Annual Meeting of the American Anthropology Association, where we presented "The Bioarchaeology of Identity: Case Studies from the South Central Andes." We thank the members of that symposium and its organizer for stimulating our examination of bioarchaeology and identity.

References Cited

Aldenderfer, Mark S. 1993. *Domestic Architecture, Ethnicity, and Complementarity in the South-Central Andes*. Iowa City: University of Iowa Press.

Arnold, Bettina, and Nancy Wicker, editors. 2001. *Gender and the Archaeology of Death*. Walnut Creek, Calif.: AltaMira Press.

Barth, Fredrik, editor. 1969. *Ethnic Groups and Boundaries: The Social Organization of Culture Difference*. Boston: Little, Brown and Company.

Bauer, Brian. 1996. "Legitimization of the State in Inca Myth and Ritual." *American Anthropologist* 98: 327–337.

———. 2004. *Ancient Cuzco: Heartland of the Inca*. Austin: University of Texas Press.

Bauer, Brian S., and Charles Stanish. 2001. *Ritual and Pilgrimage in the Ancient Andes: The Islands of the Sun and the Moon*. Austin: University of Texas Press.

Bentley, G. Carter. 1987. "Ethnicity and Practice." *Comparative Studies in Society and History* 29: 24–55.

Bentley, R. Alexander. 2006. "Strontium Isotopes from the Earth to the Archaeological Skeleton: A Review." *Journal of Archaeological Method and Theory* 13: 135–187.

Blom, Deborah E. 1999. "Tiwanaku Regional Interaction and Social Identity: A Bioarchaeological Approach." Doctoral dissertation, University of Chicago, Department of Anthropology.

———. 2005a. "A Bioarchaeological Approach to Tiwanaku Group Dynamics." In *Us and Them: Archaeology and Ethnicity in the Andes*, edited by R. M. Reycraft, 153–182. Los Angeles: Cotsen Institute of Archaeology, University of California at Los Angeles.

———. 2005b. "Embodying Borders: Human Body Modification and Diversity in Tiwanaku Society." *Journal of Anthropological Archaeology* 24: 1–24.

Blom, Deborah E., Benedikt Hallgrímsson, Linda Keng, María C. Lozada Cerna, and Jane E. Buikstra. 1998. "Tiwanaku 'Colonization': Bioarchaeological Implications for Migration in the Moquegua Valley, Peru." *World Archaeology* 30: 238–261.

Blom, Deborah E., and John W. Janusek. 2004. "Making Place: Humans as Dedications in Tiwanaku." *World Archaeology* 36: 123–141.

Bourdieu, Pierre. 1990. *The Logic of Practice*. Stanford: Stanford University Press.

Bowser, Brenda J. 2000. "From Pottery to Politics: An Ethnoarchaeological Study of Political Factionalism, Ethnicity, and Domestic Pottery Style in the Ecuadorian Amazon." *Journal of Archaeological Method and Theory* 7: 219–248.

Bravo, Leandro, and Agustín Llagostera. 1986. "Solcor-3: Un aporte al conocimiento de la cultura San Pedro. Período 500 al 900 d.C." *Chungará* 16–17: 323–332.

Brown, James A., editor. 1971. *Approaches to the Social Dimensions of Mortuary Practices*. Washington, D.C.: Society for American Archaeology.

Carr, Christopher. 1995. "Mortuary Practices: Their Social, Philosophical-Religious, Circumstantial and Physical Determinants." *Journal of Archaeological Method and Theory* 2: 105–200.

Chesson, Meredith S. 2001. *Social Memory, Identity and Death: Anthropological Perspectives on Mortuary Rituals*. Archaeological Papers of the American Anthropological Association No. 10. Arlington, Va.: American Anthropological Association.

Conkey, Margaret W., and Christine A. Hastorf, editors. 1990. *The Uses of Style in Archaeology*. Cambridge: Cambridge University Press.

Costa-Junqueira, María Antonietta, and Agustín Llagostera. 1994. "Coyo-3: Momentos finales del período medio en San Pedro de Atacama." *Estudios Atacameños* 11: 73–107.

Costa-Junqueira, María Antonietta, Walter Alves Neves, Ana María de Barros, and Rafael Bartolomucci. 1998. "Trauma y estrés en poblaciones prehistóricas de San Pedro de Atacama, norte de Chile." *Chungará* 30: 65–74.

D'Altroy, Terence N. 1992. *Provincial Power in the Inka Empire*. Washington, D.C.: Smithsonian Institution Press.

———. 2002. *The Incas*. Malden, Mass.: Blackwell Publishers.

D'Altroy, Terence N., and Katharina Schreiber. 2004. "Andean Empires." In *Andean Archaeology*, edited by H. Silverman, 255–279. Malden, Mass.: Blackwell Publishers.

Díaz-Andreu, Margarita, Sam Lucy, Staša Babić, and David N. Edwards, editors. 2005. *The Archaeology of Identity: Approaches to Gender, Age, Status, Ethnicity, and Religion*. London: Routledge.

Dietler, Michael, and Ingrid Herbich. 1998. "*Habitus*, Techniques, Style: An Integrated Approach to the Social Understanding of Material Culture and Boundaries." In *The Archaeology of Social Boundaries*, edited by M. T. Stark, 232–263. Washington, D.C.: Smithsonian Institution Press.

Dillehay, Tom, editor. 1995. *Tombs for the Living: Andean Mortuary Practices*. Washington, D.C.: Dumbarton Oaks.

Epperson, Terrence W. 1999. "The Contested Commons: Archaeologies of Race, Repression and Resistance in New York City." In *Historical Archaeologies of Capitalism*, edited by M. P. Leone and P. B. Potter, 81–110. New York: Kluwer International/Plenum Press.

Gillespie, Susan D. 2001. "Personhood, Agency, and Mortuary Ritual: A Case Study from the Ancient Maya." *Journal of Anthropological Archaeology* 20: 73–112.

Goldstein, Paul. 1993. "Tiwanaku Temples and State Expansion: A Tiwanaku Sunken-Court Temple in Moquegua, Peru." *Latin American Antiquity* 4: 22–47.

———. 2005. *Andean Diaspora: The Tiwanaku Colonies and the Origins of South America Empire*. Gainesville: University Press of Florida.

Gosden, Chris. 2004. *Archaeology and Colonialism: Cultural Contact from 5000 BC to the Present*. Cambridge: Cambridge University Press.

Grove, Matthew J., Paul A. Baker, Scott L. Cross, Catherine A. Rigsby, and Geoffrey O. Seltzer. 2003. "Application of Strontium Isotopes to Understanding the Hydrology and Palaeohydrology of the Altiplano, Bolivia-Peru." *Palaeogeography, Palaeoclimatology, Palaeoecology* 194: 281–297.

Hawkesworth, Christopher J., M. Hammill, A. R. Gledhill, P. van Calsteren, and Graeme Rogers. 1982. "Isotope and Trace Element Evidence for Late-Stage Intra-Crustal Melting in the High Andes." *Earth and Planetary Science Letters* 58: 240–254.

Hegmon, Michelle. 1998. "Technology, Style, and Social Practices: Archaeological Approaches." In *The Archaeology of Social Boundaries*, edited by M. T. Stark, 264–279. Washington, D.C.: Smithsonian Institution Press.

Helms, Mary W. 1993. *Craft and the Kingly Ideal: Art, Trade and Power.* Austin: University of Texas Press.

Hoshower, Lisa M., Jane E. Buikstra, Paul S. Goldstein, and Ann D. Webster. 1995. "Artificial Cranial Deformation at the Omo M10 Site: A Tiwanaku Complex from the Moquegua Valley, Peru." *Latin American Antiquity* 6: 145–164.

Hutchinson, John, and Anthony D. Smith, editors. 1996. *Ethnicity.* Oxford: Oxford University Press.

James, David E. 1982. "A Combined O, Sr, Nd, and Pb Isotopic and Trace Element Study of Crustal Contamination in Central Andean Lavas, I. Local Geochemical Variations." *Earth and Planetary Science Letters* 57: 47–62.

Jones, Siân. 1997. *The Archaeology of Ethnicity: Constructing Identities in the Past and Present.* London and New York: Routledge.

Knudson, Kelly J. 2004. "Tiwanaku Residential Mobility in the South Central Andes: Identifying Archaeological Human Migration through Strontium Isotope Analysis." Doctoral dissertation, University of Wisconsin at Madison, Department of Anthropology.

Knudson, Kelly J., and T. Douglas Price. 2007. "The Utility of Multiple Chemical Techniques in Archaeological Residential Mobility Studies: Case Studies from Tiwanaku- and Chiribaya-Affiliated Sites in the Andes." *American Journal of Physical Anthropology* 132: 25–39.

Knudson, Kelly J., T. Douglas Price, Jane E. Buikstra, and Deborah E. Blom. 2004. "The Use of Strontium Isotope Analysis to Investigate Tiwanaku Migration and Mortuary Ritual in Bolivia and Peru." *Archaeometry* 46: 5–18.

Knudson, Kelly J., Tiffiny Tung, Kenneth C. Nystrom, T. Douglas Price, and Paul Fullagar. 2005. "The Origin of the Juch'uypampa Cave Mummies: Strontium Isotope Analysis of Archaeological Human Remains from Bolivia." *Journal of Archaeological Science* 32: 903–913.

Kolata, Alan L. 1993. *The Tiwanaku: Portrait of an Andean Civilization.* Oxford: Blackwell.

———. 1996. "Principles of Authority in the Native Andean State." *Journal of the Steward Anthropological Society* 24: 61–84.

———. 2003. "The Social Production of Tiwanaku: Political Economy and Authority in a Native Andean State." In *Tiwanaku and Its Hinterland: Archaeology and Paleoecology of an Andean Civilization: Volume 2, Urban and Rural Archaeology,* edited by A. L. Kolata, 449–472. Washington, D.C.: Smithsonian Institution Press.

Larick, Roy. 1991. "Warriors and Blacksmiths: Mediating Ethnicity in East African Spears." *Journal of Anthropological Archaeology* 10: 299–331.

LePaige, Gustavo. 1961. "Cultura de Tiahuanaco en San Pedro de Atacama." *Anales* 1: 19–23.

Lewis, Cecil M., Jr., Jane E. Buikstra, and Anne C. Stone. 2007. "Ancient DNA and Genetic Continuity in the South Central Andes." *Latin American Antiquity* 18: 145–160.

Lewis, Cecil M., and Anne C. Stone. 2005. "MtDNA Diversity at the Archaeological Site of Chen Chen in Perú." In *Biomolecular Archaeology: Genetic Approaches to the Past:*

Proceedings from the 19th Annual Visiting Scholar Conference, edited by D. M. Reed, 47–60. Carbondale: Southern Illinois University Press.

Lightfoot, Kent. 1995. "Culture Contact Studies: Redefining the Relationship between Prehistoric and Historical Archaeology." *American Antiquity* 60: 199–217.

Llagostera, Agustín. 1996. "San Pedro de Atacama: Nodo de complementaridad reticular." In *La integración sur andina cinco siglos después*, edited by X. Albó, M. I. Anatia, J. Hidalgo, L. Núñez, A. Llagostera, M. I. Remy, and B. Revesz, 17–42. Antofogasta, Chile: Universidad Católica del Norte de Antofogasta, Chile.

Llagostera, Agustín, Constantino Manuel Torres, and María Antonietta Costa. 1988. "El complejo psicotrópico en Solcor-3 (San Pedro de Atacama)." *Estudios Atacameños* 9: 61–98.

McGuire, Randall H. 1982. "The Study of Ethnicity in Historical Archaeology." *Journal of Anthropological Archaeology* 1: 159–178.

Mujica, Elias J., Mario A. Rivera, and Thomas F. Lynch. 1983. "Proyecto de estudio sobre la complementaridad económica Tiwanaku en los valles occidentales del Centro-Sur Andino." *Chungará* 11: 85–109.

Neves, Walter A., A. M. Barros, and M. A. Costa. 1999. "Incidence and Distribution of Postcranial Fractures in the Prehistoric Population of San Pedro de Atacama, Northern Chile." *American Journal of Physical Anthropology* 109: 253–258.

Neves, Walter A., and María A. Costa. 1998. "Adult Stature and Standard of Living in the Prehistoric Atacama Desert." *Current Anthropology* 39: 278–281.

Oakland Rodman, Amy. 1992. "Textiles and Ethnicity: Tiwanaku in San Pedro de Atacama, North Chile." *Latin American Antiquity* 3: 316–340.

Owen, Bruce. 2005. "Distant Colonies and Explosive Collapse: The Two Stages of the Tiwanaku Diaspora in the Osmore Drainage." *Latin American Antiquity* 16: 45–80.

Price, T. Douglas, Clark M. Johnson, Joseph A. Ezzo, Jonathan Ericson, and James H. Burton. 1994. "Residential Mobility in the Prehistoric Southwest United States: A Preliminary Study Using Strontium Isotope Analysis." *Journal of Archaeological Science* 21: 315–330.

Price, T. Douglas, Linda Manzanilla, and William D. Middleton. 2000. "Immigration and the Ancient City of Teotihuacan in Mexico: A Study Using Strontium Isotope Ratios in Human Bone and Teeth." *Journal of Archaeological Science* 27: 903–913.

Pyszczyk, Heinz W. 1989. "Consumption and Ethnicity: An Example from the Fur Trade in Western Canada." *Journal of Anthropological Archaeology* 8: 213–249.

Rakita, Gordon F. M., Jane E. Buikstra, Lane A. Beck, and Sloan Williams, editors. 2005. *Interacting with the Dead: Perspectives on Mortuary Archaeology for the New Millennium*. Gainesville: University Press of Florida.

Rattray, Evelyn C. 1990. "The Identification of Ethnic Affiliation at the Merchant's Barrio, Teotihuacan." In *Ethnoarqueología: Coloquio Bosch-Gimpera*, edited by Y. Sugiura Y. and M. C. Serra P., 113–138. Mexico City: Universidad Nacional Autónoma de México.

Reycraft, Richard M., editor. 2005. *Us and Them: Archaeology and Ethnicity in the Andes*. Los Angeles: Cotsen Institute of Archaeology, University of California, Los Angeles.

Rogers, Graeme, and Christopher J. Hawkesworth. 1989. "A Geochemical Traverse across the North Chilean Andes: Evidence for Crust Generation from the Mantle Wedge." *Earth and Planetary Science Letters* 91: 271–285.

Romanucci-Ross, Lola, and George De Vos. 1995. *Ethic Identity: Creation, Conflict, and Accommodation.* Walnut Creek, Calif.: AltaMira Press.

Rothschild, Nan A. 2006. "Colonialism, Material Culture, and Identity in the Rio Grande and Hudson River Valleys." *International Journal of Historical Archaeology* 10: 72–107.

Sackett, James R. 1977. "The Meaning of Style in Archaeology: A General Model." *American Antiquity* 42: 369–380.

Saxe, Arthur A. 1970. "The Social Dimensions of Mortuary Practices." Doctoral dissertation, University of Michigan, Department of Anthropology.

Silverman, Helaine, and David B. Small, editors. 2002. *The Space and Place of Death.* Archeological Papers of the American Anthropological Association No. 10. Arlington, Va.: American Anthropological Association.

Smith, Stuart Tyson. 2003. *Wretched Kush: Ethnic Identities and Boundaries in Egypt's Nubian Empire.* London: Routledge.

Stahl, Ann B. 1991. "Ethnic Style and Ethnic Boundaries: A Diachronic Case Study from West-Central Ghana." *Ethnohistory* 38: 250–275.

Stanish, Charles. 2003. *Ancient Titicaca: The Evolution of Complex Society in Southern Peru and Northern Bolivia.* Berkeley: University of California Press.

Stein, Gil J. 1999. *Rethinking World Systems: Diasporas, Colonies, and Interaction in Uruk Mesopotamia.* Tucson: University of Arizona Press.

———. 2002. "From Passive Periphery to Active Agents: Emerging Perspectives in the Archaeology of Interregional Interaction." *American Anthropologist* 104: 903–916.

Stovel, Emily M. 2001. "Patrones funerarios de San Pedro de Atacama y el problema de la presencia de los contextos Tiwanaku." *Boletín de Arqueología PUCP* 5: 375–395.

Torres, Constantino M. 1985. "Estilo e iconografía Tiwanaku en las tabletas para inhalar substancias psicoactivas." *Diálogo Andino* 4: 223–245.

———. 1987. "The Iconography of the Prehispanic Snuff Trays from San Pedro de Atacama, Northern Chile." *Andean Past* 1: 191–245.

Torres, Constantino M., and William J. Conklin. 1995. "Exploring the San Pedro de Atacama/Tiwanaku Relationship." In *Andean Art: Visual Expression and Its Relation to Andean Beliefs and Values,* edited by P. Dransart, 78–108. Aldershot: Avebury.

Torres, Constantino M., David B. Repke, Kelvin Chan, Dennis McKenna, Agustín Llagostera, and Richard Evan Schultes. 1991. "Snuff Powders from Pre-Hispanic San Pedro de Atacama: Chemical and Contextual Analysis." *Current Anthropology* 32: 640–649.

Torres-Rouff, Christina. 2002. "Cranial Vault Modification and Ethnicity in Middle Horizon San Pedro de Atacama, Chile." *Current Anthropology* 43: 163–171.

Varela, Héctor Hugo, and José Alberto Cocilovo. 2000. "Structure of the Prehistoric Population of San Pedro de Atacama." *Current Anthropology* 41: 125–132.

Wells, Peter S. 1998. "Identity and Material Culture in the Later Prehistory of Central Europe." *Journal of Archaeological Research* 6: 239–298.

White, Christine D., Michael W. Spence, Fred J. Longstaffe, and Kimberly R. Law. 2004. "Demography and Ethnic Continuity in the Tlailotlacan Enclave of Teotihuacan: The Evidence from Stable Oxygen Isotopes." *Journal of Anthropological Archaeology* 123: 385–403.

White, Christine D., Rebecca Storey, Fred J. Longstaffe, and Michael W. Spence. 2004. "Immigration, Assimilation, and Status in the Ancient City of Teotihuacan: Stable Isotopic Evidence from Tlajinga 33." *Latin American Antiquity* 15: 176–198.

Wiessner, Polly. 1983. "Style and Social Information in Kalahari San Projectile Points." *American Antiquity* 48: 253–276.

10

The Bodily Expression of Ethnic Identity

Head Shaping in the Chilean Atacama

CHRISTINA TORRES-ROUFF

Investigations of body modifications allow archaeologists to discern individuals and their agency in prehistory, a crucial element in exploring societal structures. Parents in many cultures bind the heads of infants at birth to impart a permanent and socially meaningful marker of identity. Despite this, research on body modifications by physical anthropologists has taken a primarily biological perspective, emphasizing the effect of modification on particular skeletal features. Deliberate head binding can produce extreme alterations to the individual's natural head shape, yet it is unlikely that the craniometric details of the process interested those who practiced it.

The reshaping of the head is a long and intimate process of inscribing cultural identity on the body. This chapter integrates biological and archaeological information to explore the social implications of cranial vault modification in human skeletal remains from San Pedro de Atacama, Chile. Atacameños interacted with foreign powers and local exchange partners and witnessed substantial demographic shifts and environmental disruption. One of the many ways in which they acted on and reacted to these changes was by means of the permanent and culturally dictated alteration of head shape. Remains from the Quitor and Coyo phases (AD 600–1000; $n = 459$) reveal the use of this visible symbol to affiliate with foreign powers concurrent with the maintenance of bodily expressions of local identity. In contrast, 441 crania from the subsequent Solor and Catarpe phases (AD 1000–1500) demonstrate that head shaping was used to consolidate group identity. This chapter explores the ways in which radical changes in the social environment affected the patterns of cranial modification and demonstrates the utility of expanding interpretations of cranial modification beyond the biological effects of the practice into questions of social significance.

Anthropological Observations on Head Shaping

Cranial vault modification has long been a focus of investigation for anthropologists, archaeologists, and physicians. Research has historically encompassed three large categories: studies of health, descriptive or classificatory works, and analyses of cranial morphology (see also Torres-Rouff 2003; Duncan, this volume). A number of early monographs described the practice and its results in great detail and referred to cranial modifications variously as "deformations" and "ethnic mutilations" (e.g., Dembo and Imbelloni 1938; Dingwall 1931). While these research foci originated in nineteenth-century analyses of brain function, deforming techniques, and cranial morphology, technical studies are still common in contemporary investigations. A number of works have been produced in recent decades concerning the effects of modification on the elements of the skull, the complexity and shape of the sutures, and nonmetric trait frequencies (e.g., Anton et al. 1992; Cheverud and Midkiff 1992; Konigsberg et al. 1993; O'Loughlin 1996).

The social context of cranial vault modification has received surprisingly little anthropological attention over the years, although it surfaced in several ethnographic and medical reports during the late nineteenth and early twentieth centuries (e.g., Dingwall 1931; Wyman 1881–1882). More recently, it has increasingly been the focus of study for bioarchaeologists who are attempting to meld biological data with social information. For example, Vera Tiesler Blos (1998, 2001) has conducted in-depth investigations into head shaping among the Maya. Among her findings, she attributes differences within individual sites to social organization, residence patterns, and even familial practices suggesting personal intervention in head shaping rather than trained practitioners. Deborah Blom (2005) examined the use of cranial modification as a group indicator at Tiwanaku, Bolivia, neighboring sites in the Titicaca Basin, and affiliated sites in Moquegua, Peru. She documents regional patterning in cranial modification and a surprising variety of types at Tiwanaku itself, suggesting that the center of the state drew in people from nearby regions, while "outside the capital and the fluid boundary, we see a strong sense of local identity displayed symbolically with homogeneity in culturally-constructed head shape" (Blom 2005: 18). Exploring status in Halaf and Ubaid burials from northern Iraq, Theya Molleson and Stuart Campbell (1995: 50) posit that "the practice has considerable potential for elitism," which, in their case, correlates with archaeological evidence of in-cipient stratification and the concurrent emergence of well-defined social divisions. A brief perusal of these recent studies and others in the same vein suggests that the social implications of cranial vault modification are quite specific to cultural choices.

For a community, the biological nature of head shaping is secondary to its social implications. Because of the importance of cranial modification to the societies that practiced it, the diligence with which it was applied, and its frequency in antiquity, data on this subject are a fertile source of information for examining questions about social identity and social change. In this research, the analysis of cranial modification integrates contemporary approaches to the body in archaeology by attempting to unite the cultural elements of the practice with "bodily physicality" (Joyce 2005: 140). Cranial vault modification is an archaeologically visible manifestation of the merging of the physical components of the body with the expression of cultural ideas (see also Duncan, this volume). In this practice, social thought is embodied in the individual members of the group.

The Biological and Social Body

In order to interpret data concerning cranial shaping in San Pedro de Atacama, it is necessary to understand the practice as a nexus of the biological and the cultural. Numerous writers have perceived the human body not as a purely biological phenomenon but rather as malleable matter shaped and constrained by the society in which an individual participates (see also White et al., this volume). Pioneers in the anthropology of the body, including Mary Douglas, Erving Goffman, and Michel Foucault, argue that the body both is controlled by and is a component of the actions of society (Shilling 1993: 70). The construction of the body as a symbol that serves to convey social information within a culture is of particular interest in this analysis. The body in the case considered here is reshaped into a physical symbol of membership in a social community.

Throughout human history, body modifications have been used to project group membership (e.g., Brain 1979). The changes often reflect the relationships being conveyed. For example, ephemeral modifications can symbolize short-lived states of being. Investigations of native Tierra del Fuegians reveal the use of body paint to represent mood, plans, or other temporary changes in activity and social conduct (Fiore 2005; Gallardo 2004). This manipulation of the body reflects the "fluidity of embodiment" that is evident in temporary manipulations of the body's surface (Joyce 2005: 145).

In contrast, people who practice intentional cranial modification bind the heads of infants for a period of months or years to obtain a desired form. Cranial vault modification is a permanent alteration, a characteristic that is emphasized by its placement on the body of infants who cannot participate in the process. The plasticity of the skeleton is tied to age, making cranial modification something that must be imposed on infants. Additionally, biology constrains the practice by limiting potential shapes. In this sense head shaping is particularly representative of social values yet reflects biological structures.

Early in life, the modified head shape becomes a fundamental part of an individual's identity and a means of integration into the group. In many prehistoric societies, children were, in part, socialized through this common practice (see also Duncan, this volume). Despite the global presence of cranial vault modification in prehistory, consistency and homogeneity in head shape seem to be culturally specific and reflect the social weight that these modifications carry for particular groups (e.g., Brain 1979; Dingwall 1931; Tiesler Blos 2001). These altered bodies were viewed as cultural symbols by the groups that used them, which suggests that integrative, multidisciplinary perspectives on cranial vault modification may yield insight into its use in prehistory. The use of the human body to create differences and similarities in a society where they do not necessarily exist biologically is a crucial concept for understanding the use of intentional head shaping in prehistory. The very mark of the modified head is a signifier of great value. Moreover, it is one that would have been highly visible and sparked recognition or some form of understanding in members of the group and even among individuals outside of it.

The San Pedro de Atacama Oases of Northern Chile

The San Pedro oases are located in the Atacama Desert of northern Chile (figure 10.1), in an area distinguished by its extreme aridity and excellent preservation and, archaeologically, by a long and well-documented history of settlement (Latcham 1938; Llagostera 2004; Mostny 1949). The oases have been continuously occupied by groups of sedentary agro-pastoralists since approximately AD 100. Two periods of this longer history that witnessed tremendous change, the Middle Horizon and the subsequent Late Intermediate Period, are considered here.

In the Chilean Atacama, the Middle Horizon (ca. AD 500–1000) encompasses the local Quitor and Coyo phases and is considered the cultural peak

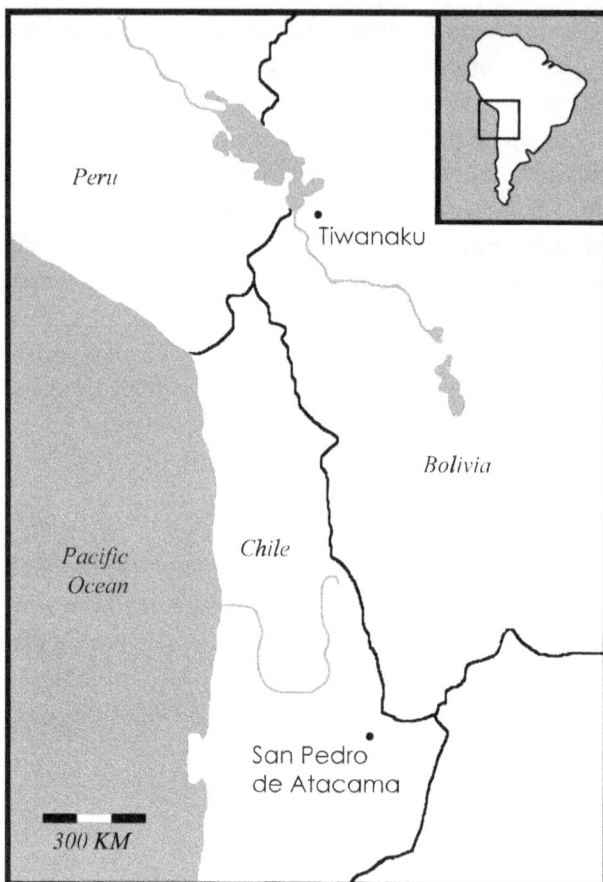

Figure 10.1. Map of the South Central Andes.

of Atacameño society (Berenguer and Dauelsberg 1989). During this prosperous time, the population grew substantially, and tombs display evidence
of well-developed status hierarchies (Berenguer and Dauelsberg 1989; Llagostera 2004; Núñez 1991). Affluence is demonstrated through the presence
of valuable metals such as gold and copper and the elaborate and finely
made ceramics and textiles interred with the dead (Oakland Rodman 1992;
Stovel 2001; Uribe and Aguero 2001). Consistent with this, bioarchaeological research has demonstrated increased stature, a sign of strong community health (Neves and Costa 1998).

During the Middle Horizon, the inhabitants of San Pedro de Atacama
interacted with the Tiwanaku state, located in the Bolivian *altiplano* (figure
10.1). There is debate as to whether Tiwanaku was physically present in the
form of representatives of the state and its authority or whether San Pedro

was enmeshed in a loose confederacy bound together by economics and Ti-wanaku ideology (Berenguer and Dauelsberg 1989; Browman 1997; Knud-son and Blom, this volume; Kolata 1993; Oakland Rodman 1992; Orellana 1985). Regardless of the form of this interaction, the presence of Tiwanaku and its goods contributed to the prestige of local elites. The state's collapse in concert with degrading environmental conditions at the end of the Middle Horizon likely affected the social hierarchy of the oases and, perhaps, the way in which these social groupings were manifested in head shape.

During the subsequent Late Intermediate Period (ca. AD 1000–1400), which includes the Yaye and Solor phases in the local chronology, changes in the settlement pattern indicate substantial population aggregation and the construction of fortified sites, suggesting the unification of groups in the Atacama for defense. Severe environmental decline coincided with these developments. Climatic shifts likely caused resource stress in the already en-vironmentally marginal societies of the Atacama Desert (Berenguer 2004: 505). These changes had significant cultural consequences. The diminished quantity and quality of grave goods in Late Intermediate Period tombs stand in particularly sharp contrast to the material wealth of the Middle Hori-zon (Berenguer 2004; Núñez 1991; Schiappacasse et al. 1989; Tarragó 1968). Given these strains, the Atacameño community may have sought unity and group cohesion during tumultuous times. The following sections analyze the way in which these radical changes in the social environment precipi-tated by environmental disruption affected the patterns of cranial modifica-tion.

A Case Study: Head Shaping in Prehistoric San Pedro de Atacama

This analysis considers the remains of 900 individuals from seven ceme-teries encompassing the Middle Horizon and Late Intermediate Period in San Pedro de Atacama (table 10.1).[1] Data collection for human remains was based on standard bioarchaeological protocols (Buikstra and Ubelaker 1994; Buzon et al. 2005; Steckel and Rose 2002). Modified crania were classed as annular, where the head is bound to produce a circumferential constric-tion of the vault, and tabular, where pressure is exerted from both the an-terior and posterior, creating a broadened appearance (figure 10.2). These two common Andean types are radically different in their appearance and would have been recognizably distinct in living populations. In addition to this classification, skulls were also examined for the angle of pressure on the posterior—either erect or oblique (figure 10.3).

Table 10.1. Patterns of Cranial Vault Modification in the Sample

Period	Site	Males			Females			Indeterminate Sex		
		Absent	Tabular	Annular	Absent	Tabular	Annular	Absent	Tabular	Annular
Middle Horizon, AD 600–1000										
	Solcor 3	26	21	5	19	18	11	7	1	0
	Tchecar Sur	61	11	9	46	19	18	23	6	0
	Coyo Oriental	29	35	3	37	46	8	0	0	0
Late Intermediate Period, AD 1000–1476										
	Quitor 6	4	2	0	6	5	0	0	4	0
	Coyo 3	9	6	1	4	11	0	1	1	0
	Catarpe	36	99	9	11	48	7	7	26	0
	Yaye	28	24	2	36	35	6	6	7	0

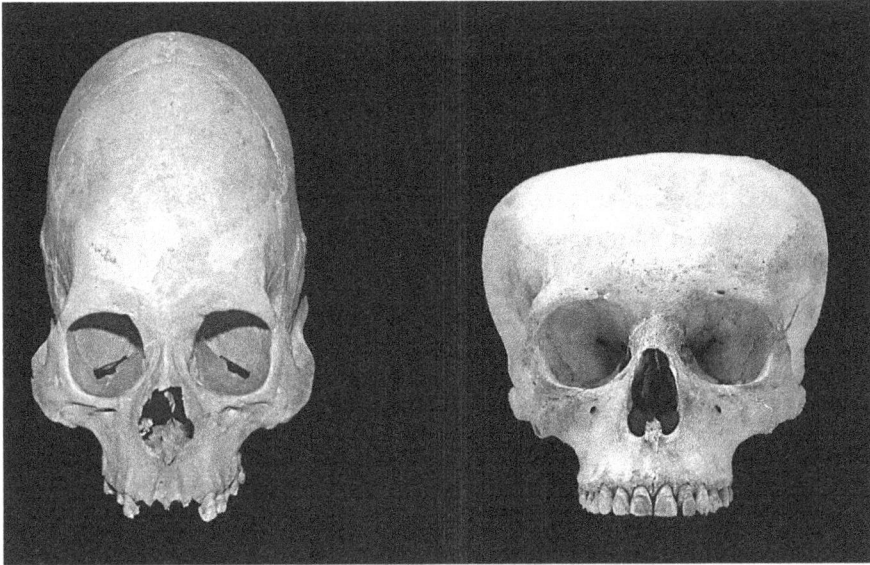

Figure 10.2. Common types of cranial vault modification in the Andes include annular modification (*left*) from Tama Tam Chullpa (Catalog #99/3195, American Museum of Natural History) and tabular modification (*right*) from Chancay E (Catalog #12-3196, Phoebe Apperson Hearst Museum of Anthropology) (photos by Christina Torres-Rouff).

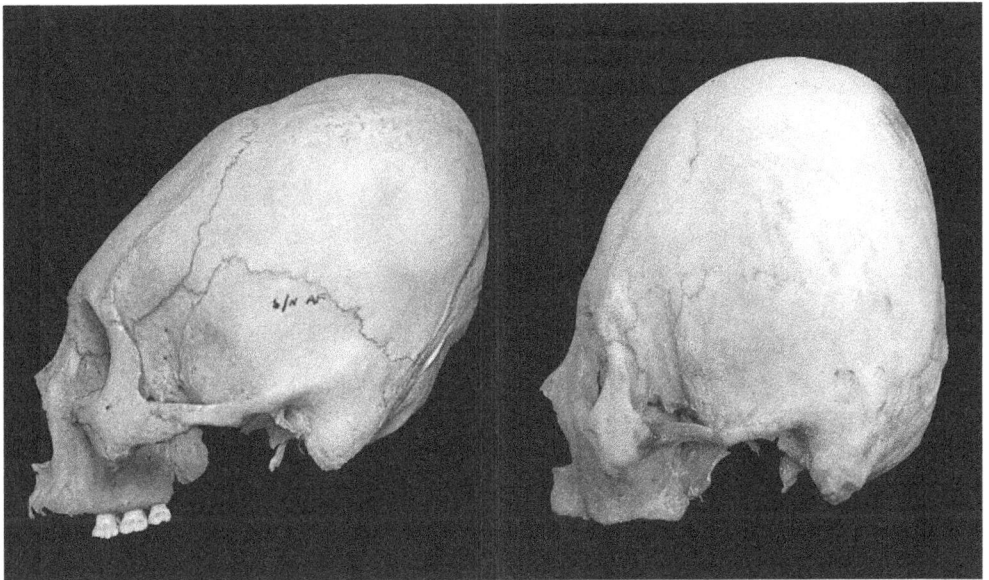

Figure 10.3. Oblique (*left*) and erect (*right*) forms of cranial vault modification from San Pedro de Atacama ([s/n], Museo Arqueológico R. P. Le Paige, San Pedro de Atacama, Chile) (photos by Christina Torres-Rouff).

While there is variation within head shapes, for this analysis modified crania are grouped into broad categories. I argue that these groupings reflect the main thrust of the modification process. First, the difference between these basic types is loosely based on the materials used to modify the head, with (tabular) or without (annular) the use of boards or directed pressure, thus giving consideration to the importance of the act of binding a child's head. Second, minor nuances in head shape that are visible in the crania were likely masked by the presence of tissue and hair in the living. Slight differences were almost certainly unintentional variations resulting from idiosyncratic physiologies and the diligence with which the modification apparatus was applied. The issues examined in this research are focused on the visibility of the modification among the living and its role within their society. Therefore a simplified classification highlights the most important visible feature of the practice in the prehistoric South Central Andes: the difference between the broadened skull and the lengthened one.

By understanding the modified head as a reflection of social values we can interpret the impact of the changes seen in the archaeological record over the periods considered here. I hypothesize that the loss of internal hier-archies and a shift toward group solidarity during times of stress resulted in the use of the body itself to present a unified group identity. A shift toward a cohesive community could manifest itself through the homogenization of patterns of deliberate head shaping.

Comparing the data from these periods (table 10.1), a statistically signifi-cant increase in the practice of cranial modification is visible ($\chi^2 = 37.386$, $df = 1$, $p = 0.0005$). The proportion of modified individuals increased from 46.0% ($n = 211/459$) to 66.4% ($n = 293/441$). Interestingly, the degree to which an individual is modified shows little consistency, with a range of moderate-to-pronounced modifications visible across sex, site, and time pe-riod. Moreover, changes occur not only in the frequency of this custom but also in the general type employed by the group. While tabular forms pre-dominate in the Middle Horizon, there is a substantial presence of modified individuals whose heads are shaped in an annular style ($n = 54/211$; 25.6%). In contrast, the Late Intermediate Period sees a sharp decrease in the pres-ence of annular forms ($n = 25/293$; 8.5%), while tabular types dominate the sample. This is particularly true for the tabular erect form, which is used by over 90% of the modified population in the Late Intermediate Period ($n = 268/293$). Results are consistent across the cemeteries in both periods.

These patterns persist to varying degrees when the sample is evaluated with regard to sex. Presence of modification increases in both sexes over

time (females from 54.1% to 66.3%; males from 42.0% to 65.0%). In concert with this, the presence of annular forms decreases between the periods
(females from 30.8% to 11.6%; males from 20.2% to 8.4%). In both times,
the tabular erect form is the most frequent, reaching a high in the Late Intermediate Period of 83.9% of the modified males ($n = 120/143$). A number
of more subtle differences between the sexes also exist. The most notable
of these is the higher frequency of annular forms among the females, particularly in the Middle Horizon, where it reaches statistical significance (females: 30.8%; males: 20.2%; $\chi^2 = 8.890$, $df = 2$, $p = .012$). These data suggest
that the many changes in the Atacameño social world during these periods
impacted the cultural practice of head shaping.

Discussion

Approaching the practice of head shaping through an interpretive paradigm
based on the body as an integral part of society, what can be extracted from
the documented patterns of head shaping in San Pedro de Atacama, both
between the periods considered here and as a more general characteristic
of this culture? Immediately, the lack of a strong difference based on sex is
visible; no form or type is exclusively male or female. Given that the practice is performed in infancy, this speaks to cranial modification as a child-
rearing practice that operates outside of the sexualization of the body (see
also Duncan, this volume).

Bernardo Arriaza (1988: 17) has noted, in his investigations of cranial
vault modification among the Chinchorro, that the preponderance of minor variations visible in altered head shapes indicates that the practitioner
was most likely a family member. Vera Tiesler Blos (1999: 4) takes this further, citing the similarity in modifications between males and females as an
indicator that the performance of this practice was the purview of family
members. These interpretations accord with the range of variation found in
the San Pedro sample, where each type of modification entails a continuum
from subtler to more extreme. Similarly, there are a number of individuals
whose cranial shape is not symmetrical, also signaling a practitioner who
was not a specialist within this society. It is worth emphasizing the possibility that molding the head is tied to child-rearing in the Atacama. The greater
variation visible among females in this sample could be interpreted as a
reflection of female exogamy, initially postulated for San Pedro de Atacama
based on analyses of cranial metrics (Costa and Llagostera 1994), not necessarily differential treatment for infant girls.

The higher rate of annular forms seen in the Middle Horizon sample may reflect the greater mobility and diversity that accompanies prosperous times. Annular forms are not common in the Atacama, although they are a common type in the Bolivian highlands and on the Chilean coast (Arriaza 1988; Cocilovo 1994; Marroquin 1944). Deborah Blom's (2005) research on diversity in the Tiwanaku core indicates the possibility that the influence of the state together with the affluence of the time may have resulted in an increased flow of people throughout the South Central Andes. Consequently, this may suggest a similar increase in head shape diversity in Middle Horizon San Pedro de Atacama, a pattern that was also suggested by a small craniometric study demonstrating high levels of genetic diversity in the Middle Horizon, with a subsequent decrease in admixture in later periods (Varela and Cocilovo 2000).

Annular types are seen in substantially higher numbers in the Middle Horizon sample from San Pedro de Atacama. The highest rate of circumferential head shaping is seen at Tchecar, where 42.9% of the modified population employed this type ($n = 27/63$), while it represents only 18.2% of the modified population in the other Middle Horizon cemeteries. It is particularly interesting that Tchecar displays less material wealth and fewer foreign artifacts than Coyo Oriental and Solcor 3 despite the greater variety in head shape found there. Individuals in San Pedro might have used cranial vault modification together with material indicators as signifiers of alliance with the distant Tiwanaku state, thereby conferring prestige and authority on certain individuals (Helms 1992: 161; Knudson and Blom, this volume; Torres-Rouff 2002).

The following period sees more homogenous cranial modification styles in San Pedro de Atacama that are perhaps a sign of decreased mobility in the Late Intermediate Period. Increasing homogeneity in head shape may reflect a desire to blur individual distinctions, as would be consistent with a time of flattened hierarchies and less mobility. Particularly notable is the resurgence of tabular types in the Late Intermediate Period, especially the tabular erect variant that is considered a local form for the Atacameños throughout prehistory (figure 10.4) (Latcham 1938; Munizaga 1969; Torres-Rouff 2003). The return to unified cultural norms in a highly symbolic practice such as head shaping stresses the collective nature of the group (not individual claims and status), which may have been necessary for survival and group cohesion in the turbulent Late Intermediate Period. The homogenization of head shape mirrors other archaeological indicators of Atacameño social consolidation during this time, including aggregated settlements. The creation of a unified

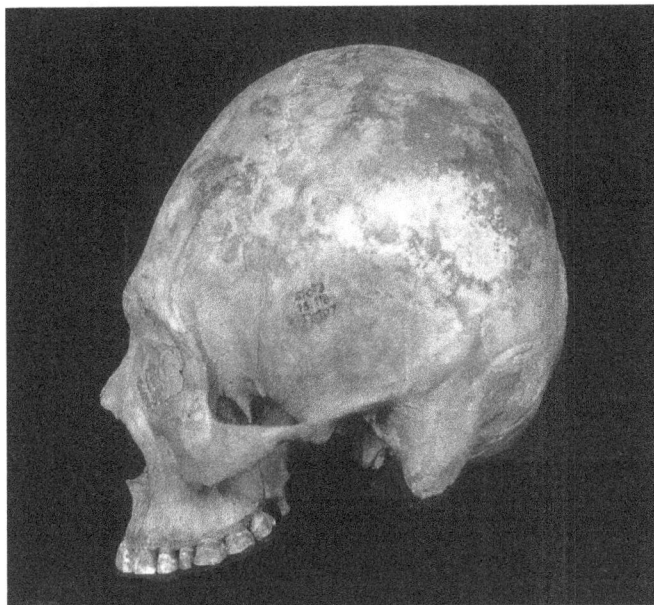

Figure 10.4. Tabular erect cranial vault modification from Solcor 3 (Catalog #13,118, Museo Arqueológico R. P. Le Paige, San Pedro de Atacama, Chile) (photo by Christina Torres-Rouff).

social identity that could be read and understood by both members of the group and outsiders was a tangible result of the reconfiguration associated with the Late Intermediate Period.

Conclusion

The archaeological body is both a physical thing and a means by which identity is represented (Rautman and Talalay 2000). To follow Lynn Meskell's (1998: 145) framework of historical approaches to the body, it is both a scene of display and an artifact. Through analyses of cranial shaping, however, we can incorporate both of these perspectives and move beyond them to an interpretation of the individual lived body.

Modified bodies can be read as symbols within their society and as such are manipulable and can react over time. The change in patterns of head shaping between the Middle Horizon and Late Intermediate Period implies that an ideological shift was concurrent with the evident external changes. Blom (2005: 2) notes that "by creating distinct differences that are not present at birth and by giving meaning to these differences, 'cultural bodies' are constructed, and symbolic boundaries created." In the case considered here, these distinctions may have been more necessary in the Middle Horizon if individuals were frequently moving between groups and exchange was more prevalent. During the Late Intermediate Period, a form of social proscription may have appeared, guiding parents to use the traditional form of modification on their children.

Exploring the social implications of head-shaping patterns in San Pedro de Atacama moves analyses of cranial modification away from consideration of the minutiae of shifts in sutural complexity and elaborate typologies of shape. These data have shown a population that reflected its internal diversity through head shape in the prosperous Middle Horizon and then shifted dramatically toward a common type during the environmental disruption and social turmoil of the subsequent Late Intermediate Period. The use of an active interpretation of body modifications allows bioarchaeologists to approach the impact of the shaped head for the populations that experienced it. Considering the social meaning of these permanent and dramatic alterations to natural head shape provides the opportunity to read more closely into the behaviors and actions of these populations to witness political alliances, societal complexity, and ethnogenesis.

Acknowledgments

I would like to thank María Antonietta Costa, Macarena Oviedo, Jessica Burns, and Krista Eckhoff for assistance in the field. This work benefited from the input of Michele Buzon and Valerie Andrushko. Portions of this research were supported by Colorado College and a Women's International Science Collaboration Grant from the American Association for the Advancement of Science.

Notes

1. All data are derived from my investigations except for the data on the Middle Horizon cemetery of Coyo Oriental ($n = 158$), where comparable data were obtained from José Cocilovo and María Zavattieri's 1994 publication.

References Cited

Anton, Susan C., Carolyn R. Jaslow, and Sharon M. Swartz. 1992. "Sutural Complexity in Artificially Deformed Human Crania." *Journal of Morphology* 214: 321–332.

Arriaza, Bernardo T. 1988. "Modelo bioarqueológico para la búsqueda y acercamiento al individuo social." *Chungará* 21: 9–30.

Berenguer, José. 2004. *Caravanas, interacción y cambio en el desierto de Atacama*. Santiago, Chile: Museo Chileno de Arte Precolombino.

Berenguer, José, and Percy Dauelsberg. 1989. "El norte grande en la órbita de Tiwanaku (400 a 1200 d.C.)." In *Culturas de Chile: Prehistoria desde sus orígenes hasta los albores de la conquista*, edited by J. Hidalgo, V. Schiappacasse, H. Niemeyer, C. Aldunate, and I. Solimano, 129–180. Santiago: Andrés Bello.

Blom, Deborah E. 2005. "Embodying Borders: Human Body Modification and Diversity in Tiwanaku Society." *Journal of Anthropological Archaeology* 24: 1–24.

Brain, Robert. 1979. *The Decorated Body*. New York: Harper and Row.

Browman, David L. 1997. "Political Institutional Factors Contributing to the Integration of the Tiwanaku State." In *Emergence and Change in Early Urban Societies*, edited by L. Manzanilla, 229–243. New York: Plenum Press.

Buikstra, Jane E., and Douglas H. Ubelaker, editors. 1994. *Standards for Data Collection from Human Skeletal Remains*. Arkansas Archeological Survey Research Series No. 44. Fayetteville: Arkansas Archeological Survey.

Buzon, Michele R., Jacqueline T. Eng, Patricia Lambert, and Philip L. Walker. 2005. "Bioarchaeological Methods." In *Handbook of Archaeological Methods*, edited by H. A. Maschner and C. A. Chippendale, 871–918. Lanham, Md.: AltaMira Press.

Cheverud, James M., and James E. Midkiff. 1992. "Effects of Fronto-Occipital Cranial Reshaping on Mandibular Form." *American Journal of Physical Anthropology* 87: 167–171.

Cocilovo, José A. 1994. "Biología de la población prehistórica de Pisagua: Continuidad y cambio biocultural en el norte de Chile." Doctoral dissertation, Universidad Nacional de Córdoba, Argentina, Facultad de Ciencias Exactas, Físicas y Naturales.

Cocilovo, José A., and María V. Zavattieri. 1994. "Biología del grupo prehistórico de Coyo Oriental (San Pedro de Atacama, Norte de Chile), II: Deformación craneana artificial." *Estudios Atacameños* 11: 135–143.

Costa, María Antonietta, and Agustín Llagostera. 1994. "Coyo-3: Momentos finales del período medio en San Pedro de Atacama." *Estudios Atacameños* 11: 73–107.

Dembo, Adolfo, and José Imbelloni. 1938. *Deformaciones intencionales del cuerpo humano de carácter étnico*. Buenos Aires: J. Anesi.

Dingwall, Eric. 1931. *Artificial Cranial Deformation: A Contribution to the Study of Ethnic Mutilations*. London: J. Bale Sons and Danielsson.

Fiore, Dánae. 2005. "Pinturas corporales en el fin del mundo: Una introducción al arte visual Selk'nam y Yamana." *Chungará* 37: 109–127.

Gallardo, Francisco. 2004. "The Perfect Image of Self and Its Scorch Mark." In *Twelve Perspectives on Selknam, Yahgan, and Kawesqar*, edited by C. Odone and P. Maso, 75–101. Santiago, Chile: Taller Experimental Cuerpos Pintados.

Helms, Mary. 1992. "Long-Distance Contacts, Elite Aspirations, and the Age of Discovery in Cosmological Context." In *Resources, Power, and Interregional Interaction*, edited by E. M. Schortman and P. A. Urban, 157–174. New York: Plenum Press.

Joyce, Rosemary. 2005. "Archaeology of the Body." *Annual Review of Anthropology* 34: 139–158.

Kolata, Alan L. 1993. *The Tiwanaku: Portrait of an Andean Civilization.* The Peoples of America. Cambridge, Mass.: Blackwell.

Konigsberg, Lyle W., Luci A. P. Kohn, and James M. Cheverud. 1993. "Cranial Deformation and Nonmetric Trait Variation." *American Journal of Physical Anthropology* 90: 35–48.

Latcham, Richard E. 1938. *Arqueología de la región atacameña.* Santiago: Universidad de Chile.

Llagostera, Agustín. 2004. *Los antiguos habitantes del Salar de Atacama: Prehistoria Atacameña.* Santiago: Pehuen.

Marroquin, José. 1944. "El cráneo deformado de los antiguos aimaras." *Revista del Museo Nacional* 13: 15–40.

Meskell, Lynn. 1998. "The Irresistible Body and the Seduction of Archaeology." In *Changing Bodies, Changing Meanings: Studies on the Human Body in Antiquity,* edited by D. Montserrat, 139–161. London and New York: Routledge.

Molleson, Theya, and Stuart Campbell. 1995. "Deformed Skulls at Tell Arpachiyah: The Social Context." In *The Archaeology of Death in the Ancient Near East,* edited by S. Campbell and A. Green, 45–55. Oxford: Oxbow Monographs.

Mostny, Greta. 1949. "Ciudades atacameñas." *Boletín del Museo Nacional de Historia Natural (Santiago de Chile)* 24: 125–211.

Munizaga, Juan. 1969. "Deformación craneana intencional en San Pedro de Atacama." In *Actas del V Congreso Nacional de Arqueología Chilena,* 129–134. La Serena, Chile: Museo Arqueológico de La Serena.

Neves, Walter A., and María Antonietta Costa. 1998. "Adult Stature and Standard of Living in the Prehistoric Atacama Desert." *Current Anthropology* 39: 278–281.

Núñez, Lautaro. 1991. *Cultura y conflicto en los oasis de San Pedro de Atacama.* Santiago: Editorial Universitaria.

Oakland Rodman, Amy. 1992. "Textiles and Ethnicity: Tiwanaku in San Pedro de Atacama." *Latin American Antiquity* 3: 316–340.

O'Loughlin, Valerie D. 1996. "Comparative Endocranial Vascular Changes Due to Craniosynostosis and Artificial Cranial Deformation." *American Journal of Physical Anthropology* 101: 369–385.

Orellana, Mario. 1985. "Relaciones culturales entre Tiwanaku y San Pedro de Atacama." *Diálogo Andino* 4: 247–257.

Rautman, Alison E., and Lauren E. Talalay. 2000. "Introduction: Diverse Approaches to the Study of Gender in Archaeology." In *Reading the Body: Representations and Remains in the Archaeological Record,* edited by A. E. Rautman, 1–12. Philadelphia: University of Pennsylvania Press.

Schiappacasse, Virgilio, Victoria Castro, and Hans Niemeyer. 1989. "Los desarrollos regionales en el norte grande." In *Culturas de Chile: Prehistoria desde sus orígenes hasta los albores de la conquista,* edited by J. Hidalgo, V. Schiappacasse, H. Niemeyer, C. Aldunate, and I. Solimano, 181–220. Santiago: Andrés Bello.

Shilling, Chris. 1993. *The Body and Social Theory*. London/Newbury Park, Calif.: Sage Publications.

Steckel, Richard H., and Jerome C. Rose. 2002. *The Backbone of History: Health and Nutrition in the Western Hemisphere*. Cambridge: Cambridge University Press.

Stovel, Emily. 2001. "Patrones funerarios en San Pedro de Atacama y el problema de la presencia de los contextos Tiwanaku." *Boletín de Arqueología* 5: 375–395.

Tarragó, Miriam. 1968. "Secuencias culturales de la etapa Agroalfarera de San Pedro de Atacama (Chile)." *Actas y Memorias del XXXVII Congreso Internacional de Americanistas* (Buenos Aires) 2: 119–145.

Tiesler Blos, Vera. 1998. *La costumbre de la deformación cefálica entre los antiguos mayas: Aspectos morfológicos y culturales*. Mexico City: INAH.

———. 1999. "Head Shaping and Dental Decoration among the Ancient Maya: Archaeological and Cultural Aspects." www.mesoweb.com/features/tiesler/media/headshaping.pdf (accessed November 1, 2006).

———. 2001. "Head Shaping and Dental Decoration: Two Biocultural Attributes of Cultural Integration and Social Distinction among the Ancient Maya." *American Journal of Physical Anthropology Supplement* 32: 149.

Torres-Rouff, Christina. 2002. "Cranial Vault Modification and Ethnicity in Middle Horizon San Pedro de Atacama, Chile." *Current Anthropology* 43: 163–171.

———. 2003. "Shaping Identity: Cranial Vault Modification in the Pre-Columbian Andes." Doctoral dissertation, University of California, Santa Barbara, Department of Anthropology.

Uribe, Mauricio, and Carolina Aguero. 2001. "Alfarería, textiles y la integración del norte grande de Chile a Tiwanaku." *Boletín de Arqueología PUCP* 5: 397–426.

Varela, Héctor Hugo, and José Cocilovo. 2000. "Structure of the Prehistoric Population of San Pedro de Atacama." *Current Anthropology* 41: 125–132.

Wyman, Hal C. 1881–1882. "Artificial Deformity of the Skull and Unsoundness of Mind." *Detroit Lancet* 5: 428–430.

PART III

Concluding Remarks

11

Identity Formation

Communities and Individuals

JANE E. BUIKSTRA

While many archaeologists and bioarchaeologists recognize the multifactorial nature of identity, in practice most studies emphasize only one or at most two dimensions of variability, commonly age and gender or gender and ethnicity. This is seen in the Gowland and Knüsel (2006) volume and in various archaeological treatments (e.g., Díaz-Andreu et al. 2005; Insoll 2007). As Knudson and Stojanowski emphasize in chapter 1, this volume departs from this model and heeds Meskell's (2007) clarion call for the integration of multiple identities.

The two sections of this volume, one centering on group or community identities and the other focusing on individuals, have intellectual roots in two distinct traditions of the study of human remains from archaeological sites. The first can be traced to the population-based bioarchaeology that was initially influenced by the "new archaeology" begun in the 1960s (Buikstra 1977; see also Buikstra 2006: 348–353 for a discussion of "the bioarchaeologies"). The second, individual emphasis develops from Saul's (1972; Saul and Saul 1989) osteobiographic method, initially directed to Maya contexts, and is more forensic in nature. While both precursor approaches clearly were concerned with archaeological contexts, their treatments were not so nuanced or so commonly enriched by ethnographic or ethnohistorical documentation as these descendants.

The Bioarchaeology of Identity

In chapter 1 of this bioarchaeological study of identity, Knudson and Stojanowski (this volume) make the following important distinction between this volume and many prior examples of identity studies:

> Rather, we define identities research as not about who people were or where they or their ancestors came from but about who they thought they were, how they advertised this identity to others, how others per-

ceived it, and the resulting repercussions of this matrix of interpersonal and intersocietal relationships. More specifically, the questions that bioarchaeology addresses are as follows. How do identities begin and manifest both at the level of the individual, as in ensoulment and the creation of personal identity, and at the level of the community, as in ethnogenesis, ethnic emergence, and coalescence? How are markers of identity overtly displayed and manipulated across time and space, as in cranial modification or the use of specific material culture styles? How does the presence of multiple social identities, or social plurality, manifest itself? In what contexts does plurality lead to health disparity and under what circumstances is plurality less deleterious?

Knudson and Stojanowski (this volume) thus set a high standard for the participants in this volume and for subsequent bioarchaeological identity studies, especially those of ethnogenesis, identity transcendence, embodiment, and formations of social, political, and religious identities, which are the foci here.

One of the measures of a volume's success is whether it meets its stated goals. As Knudson and Stojanowski explicitly state in chapter 1 (this volume), these are four in number (emphasis added):

First, we explicitly *minimize the presentation of data and methodologies* in an attempt to shed bioarchaeology's status as a methodological tool. Second, chapter contributors present data sets and analyses generated through well-established bioarchaeological techniques but *push their interpretations* into new and sometimes challenging places. The importance of this volume lies in the new ways in which authors *link their bioarchaeological data to broader problem orientations within the social sciences.* Third, the chapters highlight the *unique benefits of bioarchaeological data sets* and present novel approaches to the past that are not available using other data sources. Fourth, and finally, the chapter contributors stress the importance of *using multiple lines of evidence* to effect more synthetic and nuanced interpretations of social identities in the archaeological record.

The authors have succeeded in meeting these goals. We see that bioarchaeology is indeed useful in addressing contemporary issues, such as the impact of colonization upon ethnogenesis and identity formation in chapters 3, 4, 6, 9, and 10. In parallel, identity development in the face of political upheaval and state collapse is addressed by Sutter (this volume) and Torres-

Rouff (this volume). The relation of local to global processes is impressive, as is the link to general theories of state formation and collapse.

All the chapters utilize multiple lines of evidence, melding distinctive sources such as ethnography and ethnohistory with nuanced consideration of archaeological contexts. Especially notable is Stojanowski's (this volume) identification of a "Spanish Indian" identity and ancestry for the present-day Seminole, with implications for their cultural patrimony. Despite Siân Jones' (1997) emphasis upon the fluid nature of ethnic identity, as discussed above, Stojanowski's argument concerning a temporal depth for Seminole ancestors in or near their present homelands is clear and convincing. Importantly, Stojanowski's results uniquely address the ancestry of the Seminole and run counter to historical sources (or lack thereof) and oral traditions. As Stojanowski (this volume) also notes, this is also the first time that ethnogenesis has been documented through the study of biodistance data.

Impressive, too, is Knudson and Blom's (this volume) masterful melding of multiple lines of bioarchaeological data, including bone chemistry, biodistance, cranial modification, cemetery structure, and grave contents, to address political, social, and religious identity maintenance and transformation in the face of Tiwanaku influence far beyond the heartland in the southern Peru site of Chen Chen and the northern Chilean *altiplano* site of San Pedro de Atacama.

A further creative study is Duncan's (this volume) proposal that ancient Maya without cranial modification were no less embodied than those whose heads were artificially shaped. Employing an ethnographic model to argue that head shaping reflects attempts to minimize soul loss during childhood rites of passage flies in the face of other traditional interpretations that emphasize ethnic or kin formations. Duncan's challenge will be to develop tests of this hypothesis through other lines of evidence. White et al. (this volume), Knudson and Blom (this volume), and Torres-Rouff (this volume) follow a more traditional model, linking modification forms to group membership.

Also countering contemporary wisdom is Sutter's (this volume) two-stage diaspora model of colonization for the Late Intermediate Chiribaya polity in southern Peru. Perhaps methodological differences explain the different interpretations of identity structure by Sutter (this volume) and Lozada and Buikstra (2002), who favor a more stable coastal polity formation.

Klaus and Tam Chang (this volume) masterfully combine ethnohistoric sources along with their detailed study of funerary ritual to argue for the transcendence of ethnic identity beyond the Spanish incursion and even its stability in the face of earlier, prehistoric polity collapses. This runs counter to social scientists' usual emphasis on the dynamic nature of ethnic identity. In future work, these authors may wish to address the possibility that, while symbol systems remained stable, their meanings may have been transformed.

Nystrom's (this volume) association of Chachapoya ethnogenesis with the impact of Inka colonization is a perfect example of the power of bioarchaeological research. Nystrom's theoretical discussion of the nature of ethnicity and ethnic formations is masterful, as is that of Stojanowski (this volume). They serve as important models for theoretical perspectives in other bioarchaeological studies of ethnicity and ethnogenesis. Similarly, White et al.'s (this volume) detailed discussion of embodiment is important, as is their emphasis on the conscious and unconscious manners by which identity may be registered upon and within the human body. Similarly, their call for a shift from an osteobiographic approach to social biography is good advice indeed.

Torres-Rouff's (this volume) use of cranial modification data, along with archaeological contextual data, to establish the nature of identity formation during and after the distant Tiwanaku state collapsed is an impressive example of the use of multiple lines of bioarchaeological evidence from a large and complex data set. Her extensive discussion of the anthropology of cranial modification is also a useful contribution.

Torres-Rouff (this volume), like Knudson and Blom (this volume), Sutter (this volume), Stojanowski (this volume), and Nystrom (this volume), illustrates the power of large, chronologically and/or regionally diverse data sets gathered over years, frequently in collaboration with other scholars. Such approaches are absolutely necessary if we are to continue to develop nuanced models for change that inform both the past and present, with implications for models of future processes.

Thus this volume, in its novel consideration of the bioarchaeology of identity in the Americas, is an important contribution to scholarly inquiry. The authors and especially the volume editors are to be congratulated on developing a thematically coherent volume that is much more than simply a collection of papers. They have indeed met their stated goals in a most impressive and coherent manner.

References Cited

Buikstra, Jane E. 1977. "Biocultural Dimensions of Archaeological Study: A Regional Perspective." In *Biocultural Adaptation in Prehistoric America*, edited by R. L. Blakely, 67–84. Athens: University of Georgia Press.

———. 2006. "Introduction to Section III: On to the 21st Century." In *Bioarchaeology: The Contextual Analysis of Human Remains*, edited by J. E. Buikstra and L. A. Beck, 347–357. New York: Academic Press.

Díaz-Andreu, Margarita, Sam Lucy, Staša Babić, and David N. Edwards. 2005. *The Archaeology of Identity: Approaches to Gender, Age, Status, Ethnicity, and Religion.* London: Routledge.

Gowland, Rebecca, and Christopher Knüsel, editors. 2006. *Social Archaeology of Funerary Remains.* Oxford: Alden Press.

Insoll, Timothy. 2007. "Introduction: Configuring Identities in Archaeology." In *The Archaeology of Identities: A Reader*, edited by T. Insoll, 1–19. London: Routledge.

Jones, Siân. 1997. *The Archaeology of Ethnicity: Constructing Identities in the Past and the Present.* London and New York: Routledge.

Lozada C., María Cecilia, and Jane E. Buikstra. 2002. *El Señorío de Chiribaya en la Costa Sur del Perú.* Lima, Peru: Instituto de Estudios Peruanos.

Meskell, Lynn. 2007. "Archaeologies of Identity." In *The Archaeology of Identities: A Reader*, edited by T. Insoll, 23–43. London: Routledge. [Originally published in *Archaeological Theory Today*, edited by Ian Hodder, 187–213. Cambridge: Polity, 2001.]

Saul, Frank P. 1972. *The Human Skeletal Remains of Altar de Sacrificios: An Osteobiographic Analysis.* Vol. 63(2), Memoirs of the Peabody Museum of Archaeology and Ethnology. Cambridge, Mass.: Peabody Museum of Archaeology and Ethnology, Harvard University.

Saul, Frank P., and Julie M. Saul. 1989. "Osteobiography: A Maya Example." In *Reconstruction of Life from the Skeleton*, edited by M. Y. İşcan and K. A. R. Kennedy, 287–302. New York: Alan R. Liss.

Contributors

Deborah E. Blom is associate professor in the Department of Anthropology at the University of Vermont.

Jane E. Buikstra is director of the Center for Bioarchaeological Research, School of Human Evolution and Social Change, Arizona State University.

William N. Duncan is assistant professor in the Department of Anthropology, St. John Fisher College.

Haagen D. Klaus is assistant professor in the Department of Behavioral Science, Utah Valley University.

Kelly J. Knudson is assistant professor in the Center for Bioarchaeological Research, School of Human Evolution and Social Change, Arizona State University.

Fred J. Longstaffe is Distinguished University Professor in the Department of Earth Sciences, University of Western Ontario. He also serves as the provost and vice-president of the University of Western Ontario.

Jay Maxwell is a doctoral candidate in the Department of Anatomy and Cell Biology, University of Western Ontario.

Kenneth C. Nystrom is assistant professor in the Department of Anthropology, State University of New York, New Paltz.

David M. Pendergast is curator emeritus of the Royal Ontario Museum and an honorary research fellow at the Institute of Archaeology, University College, London.

Rachel E. Scott is assistant professor in the Center for Bioarchaeological Research, School of Human Evolution and Social Change, Arizona State University.

Christopher M. Stojanowski is assistant professor in the Center for Bio-archaeological Research, School of Human Evolution and Social Change, Arizona State University.

Richard C. Sutter is associate professor in the Department of Anthropology, Indiana University–Purdue University, Fort Wayne.

Manuel E. Tam Chang is an archaeologist with the Universidad Nacional de Trujillo, Peru.

Christina Torres-Rouff is assistant professor in the Department of Anthropology, Colorado College.

Christine D. White is Canada Research Chair in Bioarchaeology and Isotopic Anthropology, Department of Anthropology, University of Western Ontario.

Index

www.ingramcontent.com/pod-product-compliance
Lightning Source LLC
Chambersburg PA
CBHW030810280326
41926CB00085B/304